U.S.
SPACE
GEAR

Crew of Shuttle flight STS-47. Seated left to right are Jerome "Jay" Apt and Curtis Brown. Standing are Jan Davis, Mark Lee, Robert Gibson, Mae Jemison, and Mamoru Mohri. The fiftieth Shuttle flight, Endeavour STS-47 claims a number of firsts that illustrate the broadening reach of the space program: Mae Jemison is the first African American female to fly in space; Mamoru Mohri is the first Japanese payload specialist, representing the National Space Development Agency of Japan; and Jan Davis and Mark Lee are the first married couple to crew a space flight. (NASA photo STS47-(S)-002)

U.S. SPACE GEAR

OUTFITTING THE ASTRONAUT

Lillian D. Kozloski

SMITHSONIAN
INSTITUTION
PRESS

WASHINGTON

Edited by Chester Zhivanos.
Production editing by Rebecca Browning.
Designed by Kathleen Sims.
Design assistance by Vicki Schwallenberg.

Library of Congress Cataloging-in-Publication Data

Kozloski, Lillian D.
    U.S. space gear : outfitting the astronaut / Lil-
lian D. Kozloski.
        p.   cm.
    Includes bibliographical references and index.
    ISBN 0-87474-459-8
    1. Space suits.   2. Astronautics—United
States—History.
    I. Title.   II. Title: US space gear.
TL1550.K68   1993
629.47′72—dc20   92-34611

British Library Cataloguing-in-Publication Data is
available.

A paperback reissue (ISBN 1-56098-382-5) of the
original cloth edition.

∞ The paper used in this publication meets the
minimum requirements of the American National
Standard for Permanence of Paper for Printed
Library Materials Z39.48-1984.

Manufactured in the United States of America.
10 9 8 7 6 5 4 3 2
05 04 03 02 01 00

To the reader: For the sake of simplicity, each chapter's endnotes are numbered in one consecutive series, regardless of whether they annotate primary text, sidebars, or bottom-of-page explanatory interpolations.

Illustrations in this volume identified as SI photos are available through the archives of the Smithsonian's National Air and Space Museum; those identified as NASA photos are from the files of the National Aeronautics and Space Administration. Negative numbers have been included wherever possible to aid future researchers.

For permission to reproduce individual illustrations appearing in this book, please correspond directly with the owners of the works as listed in the illustration captions. The Smithsonian Institution Press does not retain the right to reproduce these illustrations individually or maintain a file of addresses for photo sources.

In all conversions of quantities, metric units should be rounded off to have the same number of significant figures as the English equivalents.

Contents

Preface

The purpose of this book is to present the history of the U.S. space suit to a general audience. This volume is not to be considered a definitive work, but rather an introduction. It is limited to the technical development of the space suit in the United States, and it serves also to document the National Air and Space Museum's collection of space suits. Brief political and economic commentaries have been included to place space suit development in context, but there is enough material in each of these areas to provide the basis for entire new research projects. I hope that museum personnel who care for aerospace flight clothing collections or modern textile or costume collections, aerospace buffs, and students and teachers interested in spaceflight who wish to take home from the museum something more substantial than their own scribbled notes will find many of their questions answered in *U.S. Space Gear*. Readers with a special interest will find here detailed references to other sources.

Designers of flying clothing encountered serious difficulties before the technology of high-altitude clothing and spacewear caught up with the requirements posed by advancing capabilities of airplanes and spacecraft. Those problems and their eventual solutions are described here. The need for brevity, however, does not permit discussion of all of the constraints surrounding the funding and pace of the space program and, in particular, space suit development.

The first chapter provides the background for the development of space suits. It explains why protective clothing is needed, what happens when the human body is exposed to extremes of temperature and to lack of pressure at high altitudes, and who needs survival flight clothing. The second chapter examines partial and full pressure suits. Subsequent chapters cover the development of space suits for the manned spaceflight programs Mercury, Gemini, Apollo, Manned Orbiting Laboratory, Skylab, Apollo-Soyuz, and Shuttle and culminate in a discussion of research on hard suits. Initiated concurrently with early Apollo suits, the hard suit concept is now used in the hybrid Shuttle suits combining a hard torso with a soft fabric covering. Similar suits are being considered for Space Station. Some of the earliest concepts for space suits were derived from diving suits and medieval armor. In the 1990s, space suit designs ap-

pear to be returning to those sources of inspiration. A final chapter summarizes how the U.S. space suit program began and explores the problems encountered in the care of the National Air and Space Museum collection and some of the solutions we have developed.

Appendix 1 describes the transfer agreement between the National Aeronautics and Space Administration and the National Air and Space Museum. Appendix 2 explains the guidelines for exhibition of space artifacts on loan from the national collection. Appendix 3 gives a brief history of plastics and the resulting impact of man-made materials and superpolymers on the textile industry. Appendixes 4 and 5 list suits in the National Air and Space Museum's preservation/study collection. Appendixes 6 through 11 give vital statistics for each manned spaceflight program. The Shuttle chart (Appendix 11) is as current as possible, but, it is an ongoing project. A list of abbreviations and the bibliography are intended as quick reference sources.

Acknowledgments

Many kind and considerate persons have contributed to the publication of this book. Russell S. Colley, Robert A. Brown, and Wayne K. Galloway, pioneering space suit designers and engineers from the B. F. Goodrich Company, graciously read the manuscript and offered their comments on the discussion of early development of full pressure suits and on the Mercury and Gemini programs. Dr. George H. Kydd of the Naval Air Development Center, Warminster, Pennsylvania, helped facilitate oral history sessions and further clarification on the Mark series of pressure suits with Russell Colley. Joseph A. Ruseckas, also a pioneering space suit designer and engineer of the David Clark Company, provided helpful suggestions on the text dealing with the development of partial pressure suits, early full pressure suits, and the Mercury, Gemini, and Apollo programs. Dr. Siegfried Hansen, American pioneer in hard suit concepts, generously loaned me his personal scrapbook loaded with information on early development of the Litton series of hard suits.

Current space suit designers and engineers who also read the manuscript and supplied many useful comments include Joseph J. Kosmo of the NASA/Johnson Space Center, who advised me on hard suit development and Apollo programs through the Shuttle, and John R. Rayfield of ILC Industries and Richard C. Wilde of Hamilton

Standard, Division of United Technologies, who were most helpful in commenting on the chapters of the manned space programs (Mercury through present development of hard suit technology). British Aerospace space suit engineers Andrew Vickery, Paul Blythe, and James Hawkins, who came to study the U.S. Preservation/Study Collection of space suits, taught me about suit construction with their excellent questions and quick assimilation of the technical aspects of the space suits.

Marty Fetherstone, a summer intern at the National Air and Space Museum, brought order to the huge collection of space suit photographs. He also sorted the quagmire of space suit documentation into a usable research collection. Jon Walker, another intern, restored order to both photo and documentation collections prior to their return to the museum's archival collection. Interns Andrea Swartfager, Tom Ratcliffe, Junco Ueda, Nicole Dickerson, Sharonne Lovett, John Regino, and Richard Jin Yao and volunteer/interns James "Scott" Ewing, Dr. Gail Vivian, Barnaby Harkins, Rudy Picarde, Patricia Zuniga, Andrew Stack, Michael Dershewitz, and Erin Hammill also refined the documentation collection. Wayne Wakefield and Ken Isbell, both collection management technicians, worked on documentation and pulled together the selected bibliography from the end notes. I appreciate the many hours Linda Neuman Ezell, former curator of manned spaceflight, logged on the development of my manuscript to help create the finished product. Appreciation is also due Gregory P. Kennedy, another former curator of manned spaceflight, who traveled to many of the NASA Space Centers and space suit manufacturing companies to research and collect material for the manu-

script. Katie Schwartzstein provided excellent editorial advice on early versions of this work.

Other readers who contributed to the technical clarity of the manuscript were Adrianne Noë, curator, and Dr. Marc S. Micozzi, director, at the Armed Forces Medical Museum; Jonathan Coopersmith, historian, Texas A & M; William David Compton, contract historian, NASA History Office/JSC; and Pam Mack, historian, Clemson University. Charles Johnson, a retired physicist from the Naval Research Laboratory, who formerly supervised many flights of early atmospheric research rocket probes, assisted me with advice on the physics of the atmosphere. Glenn Sweeting, retired curator of the National Air and Space Museum's Aeronautical flight materiel collection, assisted me with the history of early flight gear, giving me additional documents and photographs of early flight clothing. Louis Raphael Purnell, former curator of manned spaceflight at the National Air and Space Museum, accessioned the first space suit artifacts from NASA in the late 1960s. He also began the space suit documentation collection. As the first curator of the space suit collection, very special thanks are due him. He patiently tutored me in space suit development, the care of the space suit collection, and management of the loan program.

The librarians at NASA and the Library of Congress supplied many leads for unusual sources. Mary Pavlovich of the National Air and Space Museum library assisted me in finding other special books, magazines, and publications. Frank Winter, curator of rocketry at the National Air and Space Museum, gave me material on many early science fiction contributions to the development of space suits. Lee Saegesser, historian of the NASA History

Office, gave me free access to the manned spaceflight files.

Many thanks to Edith Smith, an untiring National Air and Space Museum volunteer, for her efforts in polishing my prose. Thank you also to many unnamed friends, who read portions of the manuscript, helped find unique photographs, and assisted in many other ways. Finally, thanks to Vivian Vines, who turned my scrappy drafts into works of art with her magical fingers, and to Chester Zhivanos, whose painstaking editing for Smithsonian Institution Press brought to the project the order and refinements that have turned it into this book.

A special thank you to my family who listened to my story about space suits so many times and consistently supported me.

CHAPTER I

Flying Suits

The nineteenth century had hardly turned into the twentieth, when the theory of flight became a reality. The morning of December 17, 1903, dawned cold and clear at Kill Devil Hills, Kitty Hawk, North Carolina. Frigid wind velocities of 10 to 12 meters per second (22 to 27 mph) froze over the small puddles of water that filled the sand hollows, and kept Orville and Wilbur Wright indoors until around 10 A.M. When they saw the wind velocity did not die down, they decided to attempt a flight anyway. The Wright brothers deliberately chose a desolate sandy beach where their flying machine experiments would attract little public or press attention. Weather reports indicated Kitty Hawk had the sixth highest average wind in the United States and also had sufficient clear, rain-free days in the fall, when winds occasionally rose above the average.

With the toss of a coin, Orville was chosen for the first attempt of that day. He climbed into position, lying flat to give less resistance to the wind. The engine had already been warmed up to speed. Orville shifted from side to side, checking wing warp and rudder, then moved the elevator up and down. When the pilot was ready, he moved the lever over one notch to the left and slipped off the line that held the craft in place. With a tremendous flapping and snapping of the four-cylinder engine, the machine slowly began its journey down the 18.3-m (60-ft.) take-off rail into the gusting winds. After a 12.2-m (40-ft.) run, the machine lifted majestically into the air and flew for 22 seconds carrying its passenger a mere 36.5 m (120 ft.). During the world's first sustained and controlled powered flight, Orville Wright wore an ordinary business suit, a high-collared shirt, and a tie.

The United States purchased its first military plane in 1908—a Wright Flyer. The first pilots almost immediately discovered two great needs: warm clothing to protect against cold temperatures at higher altitudes and special equipment to supply oxygen. Development of specialized clothing for high-altitude flying and space exploration consumed almost seven decades.

Once war broke out in Europe in 1914, scientists in Austria, Germany, Great Britain, and France accelerated flight technology well beyond American capabilities. Typical flying clothing in World War I consisted of a fleece-lined leather jacket, gauntlets, helmet, goggles, breeches, and boots (Figure 1.1). In reaction to pleas from American scientists for a research facility comparable to that of the Europeans, the National Advisory Committee for Aeronautics (NACA) was established in 1915 to oversee the study of problems of flight, protective clothing, and flight equipment. The problems were partially recognized by the U.S. Army, which established an Aviation Clothing Board early in 1917 to supervise the development and production of flight clothing. The first U.S. Army Air Service Medical Research Laboratory was set up at Hazelhurst Field, Mineola, Long Island, New York, in 1918.

Many early adventurers failed to survive the rigors of their flights. By the mid-1920s, ascents reached altitudes of 12,000 m (40,000 ft.). Neither pilots nor high-altitude balloonists could endure the intense cold and reduced atmospheric pressure for any length of time. Over the years, different kinds of medical researchers—physiologists, chemists, physicians, and flight surgeons*—have studied the human need for adequate oxygen and pressure in high-altitude environments and now understand a good deal about why protection is required and how it may be provided.

ATMOSPHERIC PRESSURE AND PHYSIOLOGICAL EFFECTS

Earth's atmosphere, known as air, consists primarily of four non-variable gases—78.08% molecular nitrogen (N_2), 20.95% molecular oxygen (O_2), 0.93% argon (A), and 0.03% carbon dioxide (CO_2). There are trace elements of neon, krypton, hydrogen, and xenon. Because the atmosphere is turbulently mixed, its composition is more or less constant from sea level to an altitude of around 95 km (60 miles). Beyond this point, solar radiation causes chemical reactions, decompositions, and ionizations. The components of the atmosphere tend to diffuse according to their atomic weights. Hydrogen, the lightest, soon becomes the principal atmospheric element, and there is insufficient oxygen to sustain life.

Normally, the atmospheric pressure at sea level is 101.4 kPa (kilopascal) or

---

*A flight surgeon is a military physician charged with the care, maintenance, and selection of air crew and flight personnel, not necessarily someone who performs surgery.

Spalding Equipment
For the Greatest Game in the World

No. 16
Leather Safety Helmet

The past year's experiences at the war front have brought about many innovations in Aviation wear. Suggestions of value made by aviation experts have been adopted by us, and these, together with our own improvements and new ideas for aviators' comfort, protection and added efficiency of service gathered from far and wide, have given us the most satisfactory and complete stocks of outfittings, in our judgment, it is possible to get together.

No. 6
Leather Aviators' Hood

No. HH
Leather Safety Helmet
(Pat. Applied For)

No. 40
Leather Aviators' Hood

Our equipment includes wearables alike for Navy and Army aviators, including, too, Naval Coast Defense and Submarine Chaser clothing—everything up-to-the-hour, backed by the Spalding guarantee of reliability and satisfaction.

No. AAT
Two-Piece
Leather Suit

No. ATC
Spalding
Non-Sinkable Coat

No. M
One-Piece
Leather Suit

Navy men attached to the Coast Defense service, civilian yachtsmen and others assigned by the Government to the same service will find in our Coast Defense outfittings everything of the latest pattern and device for protective wear and comfort.

1.1
The needs of fledgling U.S. Army and Navy aviators during World War I instigated many novel types of flight clothing. (SI photo 88-13621, *Flying,* A. G. Spalding & Bros., Spalding Sports Worldwide, June 1918)

14.7 pounds per square inch (psi). This is equal to the weight exerted by a column of air that is 2.54 cm square (1 inch square) in cross section and located between sea level and the top of the atmosphere (Figure 1.2). As you rise through the lower atmosphere, the pressure decreases about 12% per kilometer (about 20% per mile). As pressure drops, temperature also falls. At about 10,668 m (35,000 ft.), the temperature remains a frigid –55°C (–67°F) for quite some distance. This is the isothermal layer. Below it is the troposphere, the lowest layer of the atmosphere. Nearly all weather phenomena—clouds, rain, snow, cold and warm fronts, etc.—occur at this level. At 10,668 m is the tropopause, the dividing line between the troposphere and the constant-temperature level known as the stratosphere. The stratosphere is a layer of fast jet stream winds and very little water vapor. The ozone layer in the stratosphere is at 21,336 m (70,000 ft.). At an elevation of about 60,960 m (200,000 ft.) you enter the ionosphere, where the atoms of atmospheric gas have been highly excited by solar radiation. This area consists of several regions of ionized gases, which make possible most long-distance radio communications in the shortwave bands by bouncing radio waves back to earth instead of letting them escape into space. The region above the ionosphere is the mesosphere. This region extends upward to about 644 km (400 mi.).

To function, humans need air pressure of approximately 48.3 kPa (7 psi) and oxygen pressure of at least 13 kPa (1.9 psi). At sea level, with normal atmospheric pressure, oxygen pressure is considered to be 21.2 kPa (3.08 psi). As we breathe, there is normally less pressure in our lungs than in the atmosphere. When we inhale,

• •

Where Space Begins

Space means different things to different people. The atmosphere has no sharp upper boundary, but continues to diminish with altitude. At 30 km (100,000 ft.), the pressure is only about 1% of its value at sea level. For the electronics engineer, space starts about 8 km (5 mi.) up, where air density becomes quite low and begins to act like a vacuum. For the aeronautical engineer, space begins at about 76,200 m (250,000 ft.): Air is no longer dense enough to support winged flight. The aerospace engineer considers space to be above 30,480 m (100,000 ft.), for at this elevation, the air is not considered to restrict the thrust output of a rocket engine. But for most of us, even though the atmosphere extends to about 644 km (400 mi.), space conditions begin about 5 km (3 mi.) above our heads. At 4,572 m (15,000 ft.), air pressure has dropped so much that we find it extremely difficult to function properly, and to go higher, we must take oxygen along to survive (Figure 1.3).

• •

The proportions of various gases in the Earth's atmosphere
(Art by Sternbach)

OTHER .036%   ARGON (Ar) .934%

OXYGEN (O₂) 20.946%

NITROGEN (N₂) 78.084%

The mercury barometer, or "torricellian tube," in which the column of mercury in the tube is held up by the air pressure on the surface of the mercury in the lower bowl or reservoir *(Art by Sternbach)*

Pressure

WARNING
UNSAFE
ANOXIA

BUBBLES FORM IN BLOOD

ACCLIMATIZED INDIVIDUALS

UNACCLIMATIZED INDIVIDUALS

safe region

UNLIMITED TOLERANCE LINE

OXYGEN TOXICITY
WITHIN 24 HRS

VOLUME % OXYGEN

EQUIVALENT ALTITUDE, FEET ABOVE SEA LEVEL

BAROMETRIC PRESSURE, PSI ABSOLUTE

Human tolerances to atmospheric composition and pressures *(Art by Sternbach)*

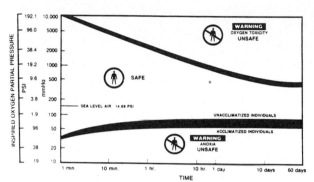

INSPIRED OXYGEN PARTIAL PRESSURE

WARNING
OXYGEN TOXICITY
UNSAFE

SAFE

SEA LEVEL AIR 14.69 PSI

UNACCLIMATIZED INDIVIDUALS

ACCLIMATIZED INDIVIDUALS

WARNING
ANOXIA
UNSAFE

TIME

Human tolerance to partial pressure of oxygen *(Art by Sternbach)*

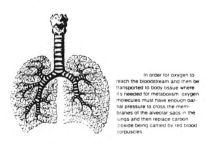

In order for oxygen to reach the bloodstream and then be transported to body tissue where it's needed for metabolism oxygen molecules must have enough partial pressure to cross the membranes of the alveolar sacs in the lungs and then replace carbon dioxide being carried by red blood corpuscles

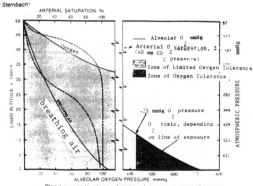

ARTERIAL SATURATION, %

Oxygen

normal breathing air

CABIN ALTITUDE × 1000 ft

ATMOSPHERIC PRESSURE

Alveolar O₂ mmHg
Arterial O₂ saturation, %
(40 mm CO₂ pressure)
Zone of Limited Oxygen Tolerance
Zone of Oxygen Tolerance

25 mmHg O₂ pressure
O₂ toxic, depending on line of exposure

ALVEOLAR OXYGEN PRESSURE mmHg

Blood oxygen levels as a function of alveolar pressure of oxygen
(Art by Sternbach)

1.2

Evangelista Torricelli (1608–47) invented the mercury barometer. He concluded that the weight of the atmosphere sustained the column of mercury inside the inverted tube. Pressures are often specified by giving the height of the mercury column at 0°C under standard gravity. This is the origin of the expression centimeters of mercury (cm/Hg) or inches of mercury (in./Hg). The pressure of earth's atmosphere at sea level will support a column of mercury 760 mm (30 in.) high, requiring 101.4 kPa or 14.7 pounds of pressure per square inch. (Art by Rick Sternbach from *The Handbook for Space Colonists*, by G. Harry Stine, copyright 1985 by G. Harry Stine, all rights reserved, SI photo 92-3971)

1.3

Altitudes reached by various space missions. (Art by Rick Sternbach from *The Handbook for Space Colonists,* by G. Harry Stine, copyright 1985 by G. Harry Stine, all rights reserved, SI photo 92-3972)

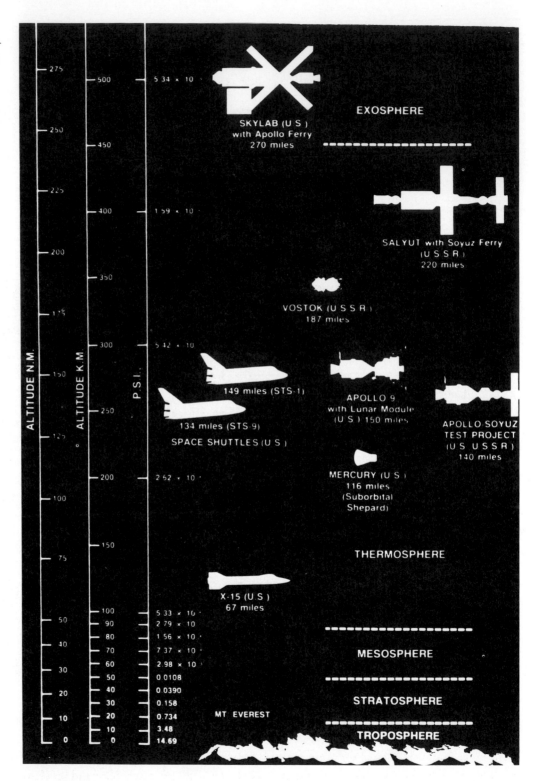

greater pressure outside the lungs forces air in with minimal muscular effort. At high altitudes, less pressure forces less oxygen in. As altitude increases, atmospheric pressure decreases exponentially. At 19,000 m (63,000 ft.), the threshold of space, air pressure is so low that nitrogen is released from body tissues and forms painful bubbles at the joints and water vapor fills the lungs. Death is imminent unless pressurized clothing is worn and pure oxygen is forced into the lungs.

French physician and physiologist Paul Bert's pressure chamber experiments in the 1860s proved that providing sufficient oxygen at higher altitudes prevents the disagreeable and dangerous effects of explosive decompression and hypoxia (oxygen deficiency in body tissues). Explosive decompression kills in two ways—by severe lack of oxygen and by air embolism. Air embolism occurs when air passages begin to close. Hypoxia results from decreased partial pressure of oxygen in inspired air and not from the mechanical effects of decreased total atmospheric pressure. In the case of hypoxia, the pilot experiences euphoria: Nothing seems to matter; the brain does not warn the body of potential dangers.[1]

Besides air pressure and oxygen needs, time is also a critical factor in avoiding hypoxia. A person breathing air has 15 seconds of useful consciousness above an altitude of 13,106 m (43,000 ft.). A pressurized suit and oxygen are necessary to control pressure that is greater inside the suit than in the atmosphere surrounding the pilot.

Researchers discovered that at about 3,000 m (10,000 ft.), oxygen pressure drops to 14.4 kPa (2.1 psi) and blood receives only 90 percent of its required oxygen supply. After two hours, a pilot may become tired and sluggish or develop a headache. At 5,500 m (18,000 ft. or 3.5 mi.), the air and oxygen pressure are half that at sea level: Oxygen pressure is down to 10.3 kPa (1.5 psi) and oxygen content in the blood is 70 percent of normal. After two hours at this altitude, a pilot can no longer make intelligent decisions, may act intoxicated, and is in danger of becoming unconscious. At 10,000 m (34,000 ft.), air pressure has dropped to less than 27.5 kPa (4 psi) and oxygen pressure is less than 5.24 kPa (0.76 psi). Less oxygen is pushed into the tissues and bloodstream. Air in the middle ear, tooth cavities, chest, and intestines expands. An oxygen mask or pressurized cabin is essential; otherwise the pilot loses consciousness within two minutes and may die.[2]

At 12,000 m (40,000 ft.), air pressure is down to 18.6 kPa (2.7 psi) and oxygen pressure is only 3.93 kPa (0.57 psi). A pilot can still survive by breathing pure oxygen but can go no higher without increasing the pressure in the cabin or by donning a pressure garment. At 15,000 m (50,000 ft. or 10 mi.), air pressure is only 11 percent of sea level pressure. The lungs fill with carbon dioxide and water vapor, since there is insufficient ambient air pressure to force in oxygen. If pressurization and pure oxygen are not available, the pilot reaches the stage of critical hypoxia (oxygen starvation) and dies within seconds. Modern aerospace researchers thus reasoned that in space, humans could breathe at 24 to 34 kPa (3.5 to 5 psi) if the surrounding gas is pure oxygen. All American manned space programs used pure oxygen inside the spacecraft until the time of Shuttle. (See Chapter 7.)

Pilots who fly at altitudes in the 15,000-to-18,000-m range (50,000-to-60,000-ft.) are in serious trouble when they experience equipment failure. If they are forced to eject or bail out of their planes without protective clothing, they risk what is called explosive decompression.

Explosive decompression was nearly the fate of Marine Lt. Colonel William H. Rankin, who lost all engine power in an F8U Crusader jet at nearly 15,000 m on July 26, 1959. The temperature inside the airplane was 21°C (70°F), but outside it was −57°C (−70°F). When Rankin bailed out, he was wearing only summer-weight flying clothing, a parachute, and his helmet. He not only remained conscious throughout most of his ordeal, but survived to give a firsthand report.

Rankin said that as soon as he ejected from the plane he felt his body become a freezing, expanding mass of pain. His abdomen stretched and bloated suddenly and violently from expanding gas. His eyes felt as if they were tearing out of their sockets. Air exploded through his ears. Cramps struck like knives over his body. Rankin described it as "savage pain." He bled from eyes, ears, nose, and mouth.

Flight surgeons agreed that if Rankin had been any higher, he would never have survived. His altitude was too close to 19,000 m* (63,000 ft.). Air pressure at this level is so low that fluids evaporate at about 36.7°C (98°F), meaning that all body fluids would vaporize. Had Rankin been 3,962 m (13,000 ft.) higher, his body would have swelled beyond recovery from gas expansion, and all vital blood vessels would have ruptured, causing death.[3]

EARLY RESEARCH AND STUDIES

Serious observation and research of the physiological effects of flying at high altitudes began in the nineteenth century. The first balloon flights had quickly made apparent the dangers of cold and altitude sickness. One important study by Paul Bert was a report on the unfortunate 1875 balloon flight of three French scientists, Joseph Crocé-Spinelli, Theodore Sivel, and Gaston Tissandier. They had experienced low-pressure breathing in Bert's steel low-atmospheric pressure chamber, but they failed to heed Bert's warning that they carried an insufficient oxygen supply for the ascent. The three men blacked out at 8,500 m (28,000 ft.). Tissandier, the lone survivor, brought his dead companions back to earth. Bert also experimented with animals using bell jars from which all air had been withdrawn (as in a vacuum chamber). His experiments showed that no matter what the atmospheric pressure was, partial pressure of oxygen at death was about 35 mm (1.4 in.) of mercury.† Humans normally live at the equivalent of 760 mm Hg (14.7 psi). This partial pressure of oxygen is the most essential atmospheric element for sustaining life.

---

*This is known as the "Armstrong Line" in honor of Brigadier General Harry G. Armstrong, who later became the second Surgeon General of the Air Force. In 1938, he published his book *Aerospace Medicine,* in which the Armstrong Line is defined. The book became the standard for the next twenty years.

---

† At 8,000 m (26,000 ft.), both the total pressure and the partial pressure of oxygen are insufficient for humans to survive for any length of time without special equipment.

Oxygen research advanced in 1901, when meteorology professors Arthur Berson and Reinhard Süring made their own balloon flight. Despite breathing pure oxygen, both passed out at 5,000 m (16,000 ft.). Süring revived and saved Berson's life by sharing his oxygen tube between them. Süring reported that survival at high altitudes did not depend on actual breathing but on oxygen saturation of the blood. He found that partial pressure of inspired oxygen decreased in proportion to decreases in atmospheric pressure until finally little or no oxygen reached the lungs. Therefore, a tightly fitted mask that forced the wearer to breathe in oxygen would be necessary above 7,924 m (26,000 ft.).[4]

The most significant early-twentieth-century work on the subject was carried out by John Scott Haldane, an English respiratory physiologist. Haldane read Paul Bert's account of the French 1875 balloon disaster and determined that some life-saving principles for balloonists pertained to deep-sea divers. A diver ascending from deep water goes from high to low pressure. The resulting effects are similar to those experienced in high-altitude flight. As a balloonist rises, the pressure against the body diminishes. For both diver and balloonist, this rapid pressure change from high to low causes gaseous nitrogen bubbles to be released from solution in the body tissues. Pain, disablement, and ultimately death result as the bubbles increase in size. Humans cannot endure such extreme changes in pressure for any length of time unless encapsulated in a pressurized suit or environment. Armored suits were first designed in 1838 to protect humans from getting caisson disease, or the "bends," caused by a too rapid ascent. In 1907 Haldane developed an oxygen pressure suit for deep-sea divers.

THE ROLE OF INDIVIDUAL PILOTS, INVENTORS, AND THE MILITARY

Pilots who sought to fly ever faster, longer, and higher and scientists who wanted to study phenomena at the upper atmosphere needed to gain regular access to altitudes above 12,000 m (40,000 ft). But during the 1930s, the military's requirement for high-flying aircraft provided the greatest incentive for the study of human endurance as it related to flight.

As early as World War I, the Germans were experimenting with the use of liquefied oxygen as a source of gaseous oxygen for high-altitude bomber pilots. Most American pilots depended on a face piece that covered the mouth and nose and fed a constant stream of oxygen through a tube. Some European and American pilots used pipe-stem oxygen delivery systems in addition to masks. American World War I pilots generally had access to helmets and clothing suitable to a summer or winter climate. After the war, U.S. Army Air Corps pilots who devoted considerable time and energy to setting new endurance records soon appreciated the need for improved flying accessories and clothing. Unfortunately, funds for the required developments were severely limited through the 1930s.[5]

Supercharged engines were developed in the early 1920s. Supercharging, or turbo charging, increases air pressure by compressing air before it reaches the intake manifold to prevent power loss at higher altitudes. Prior to this advance, pilots could not force their planes beyond 6,000 m (20,000 ft.), where warm clothing and oxygen face masks or flasks sufficed as protection. As a result of the development of new engines, planes were able to go beyond 11,000 m (36,000 ft.).

absolute. I expect to fly through rarefied areas where the pressure is as low as five pounds absolute. . . . The temperature incident with high-altitude flying will be taken care of by heating the air from the supercharger by coiling it around the exhaust manifolds."[10]

Post's first experimental suit was manufactured at B. F. Goodrich's California plant. It was constructed from six yards of double-ply rubberized parachute fabric, glued together on the bias (a diagonal direction across threads of woven material) to avoid stretching the cloth. Project engineer William R. Hucks, assisted by John A. Diehl, gave the shoulders a raglan cut and joined pigskin gloves and rubber boots to the suit. A 1.6-kg (3.5-lb.) helmet contained ear phones, a foam rubber pad on its inner roof, and a double-layer plastic visor. Post could accept food or speak through a hinged mouthpiece only when the suit was not pressurized. The helmet was fastened at both the front and back to hold it in place during pressurization. For a total cost of just under $75, B. F. Goodrich manufactured the suit for Post. Unfortunately, this first suit ruptured during unmanned pressure tests. The damaged suit was returned to the manufacturer, who sent Post to the Goodrich plant in Akron, Ohio, where he met Russell S. Colley.

Born in 1897, Colley graduated from technical college in 1918. At the B. F. Goodrich Company in Akron, Colley worked in the experimental division on the newest projects. And they were exciting ones. He developed projects as diverse as a golf ball driving machine, airplane de-icers, anti-submarine sonar, and of course, pressure suits and space suits.

Wiley Post needed a pressure suit that would hold 1.1 kg (2.5 lb.) of pressure.

Colley puzzled over what could hold that amount of pressure. He thought of a tire: 1.1 kg was nothing compared to what a tire holds. There would be a pressure-holding garment and an inner tube to hold the air. Colley developed a second suit for the pilot with a new upper and lower torso. The waist, where the first suit had ruptured, had a newly designed clamp. An oxygen hose came in just below the visor and metal rings attached the elbow and knee joints for ease of movement.

Post wore the second suit for the first time on a hot, humid July day in 1934. The suit was too tight and Post became immobilized in it. Despite the help of several strong assistants, Post could not be extricated. The design group finally led Post into a refrigerated golf ball storage room where everyone could work in comfort as they cut him out of the garment. All that remains of this suit is the helmet, which was donated to the Smithsonian in 1964.[11]

Colley took very careful measurements for Post's third suit (Figure 1.4). By August, the suit was ready for testing. Colley had made this suit of two separate layers with a large neck opening for easier entry. The outer layer was fabricated into two metal forms: one for the upper torso and arms and one for the lower torso, legs, and feet. These were dipped into liquid latex and spliced together, formed in a seated position. Gloves were cemented on, creating an airtight garment. An outer fabric suit was constructed of three-ply cotton fabric, two plies cut straight and one cut on the bias. The arms were designed so that the wearer could reach the stick and throttle. Bunching the fabric between the metal rings above and below the knee provided limited leg movement to operate the rudder pedals. These sim-

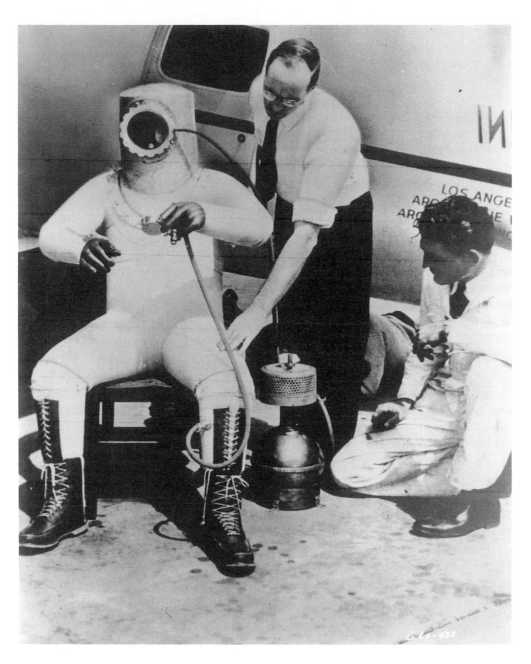

1.4
Wiley Post had his third pressurized suit in August 1934. Space suit designer Russell Colley *(middle)* checks the pressure gauge. This is believed to be the first pressurized suit worn during an actual flight. Post, flying the *Winnie Mae* on December 7, 1935, was credited by the press with establishing a new altitude record above 50,000 feet. But this was never confirmed by the National Bureau of Standards because of a malfunction of one of two required National Aeronautics Association barographs. (SI photo S81-37204)

ple adaptations to aid overall maneuverability were noteworthy accomplishments because earlier pressure suits, when pressurized, had rendered the wearer immobile.

The helmet resembled a large can with a relatively small safety-glass visor, since Post had sight in only one eye. A five-liter liquid-oxygen generator supplied pure oxygen through the helmet in a steady stream across the visor, which prevented fogging. Compressed air from the supercharger provided a mixture of air and oxygen as needed for pressurization. Near the knee, a valve with its own gauge and regulator controlled pressure for the suit. The Germans had experimented with liquid oxygen and face masks during World War I, but Wiley Post was the first to use liquid oxygen with a pressure suit.[12]

Post made his first stratospheric flight in this pressure suit in the *Winnie Mae* from Akron Municipal Airport in late August 1934. In this and succeeding flights, Post was able to slip into the jet stream. As he had suspected, this increased his flying speed from 290 to 450 km per hour (180 to 270 mph). He logged 25 hours in his tailored pressure suit, attempting to break the transcontinental speed record.

Post had proved the viability of his pressure suit and achieved stratospheric flights, despite sabotage attempts on both the suit and the airplane. On December 3, 1934, Russell Colley wrote in a report for B. F. Goodrich that during Wiley Post's most recent attempt to break the world altitude record, his pop-off valve had been jammed in a completely closed position. When he inserted the glass into the helmet porthole at about 5,500 m (18,000 ft.), pressure began to build up to nearly 48 kPa (7 psi) before Post was able to force open an emergency relief valve. In February of 1935, during another record-breaking attempt, emery dust had been added to his external supercharger which sucked the dust into the engine, ruining everything. Post glided his plane in for an emergency landing on Muroc Dry Lake. His descent was so quiet that H. E. Mertz, who was busy tinkering with his wind-powered sail car, almost passed out from fright when Post approached him. Post, garbed in his pressure suit, was merely trying to get Mertz to help him remove his helmet.

While no formal investigation ever took place, it was clearly an attempt on Post's life. Russell Colley reported that they found at least six different ways in which the plane had been sabotaged, any of which would have stopped the flight and possibly killed Wiley Post. A mechanic who had been familiar to Post's flight crew had been responsible. He did it at the request of a disgruntled pilot who felt his own chances of sponsorship were being hindered by Post's success.

Tragically, in 1935, Post died in a plane crash. His death had nothing to do with sabotage efforts. He and Will Rogers had been vacationing in Barrow, Alaska. Post was flying an experimental plane he had put together with wings, fuselage, and oversize pontoons from different sources. Russell Colley said Post made his mistake when he attempted to take off. The air was very still and there were no ripples on the water. Had he taxied around the lake several times to make it choppy, he would have made it. On that glassy, smooth water he was unable to break the suction between plane and water. When he finally did get the plane up, it rose suddenly and came straight down and crashed. Although this precluded any further immediate efforts by B. F. Goodrich to advance the development of the pressurized flying suit, Russell Colley is considered the father of the American space suit for constructing the first successful fully pressurized flying suit for Wiley Post.[13]

MILITARY NEEDS AND THE TOMATO-WORM SUIT

During the tense years of the 1930s, England, Italy, France, and Germany were all engaged in the struggle to set altitude records, and hence were also trying to develop pressure suits. The altitudes at which U.S. military aircraft operated increased approximately 3 percent each year since 1918. By 1938, the U.S. Army Air Corps recognized the utility that a flexible, reliable pressurized flying suit would have for military pilots and started its own research program. A classified project called MX-117, begun October 10, 1939, resulted in full pressure suits carrying the XH designation (X for experimental, H for high-altitude).

It was about this time that the U.S. Navy became interested in Colley's pressure suit work. At B. F. Goodrich, Carroll Krupp, Wayne Galloway, and Don Shook were assigned to a team effort with Colley to develop new pressure suits. Goodrich added other suit designers to the program as needed.

Several companies, some of which sound as surprising contenders to us today, began energetic research programs. For the most part, they were funded either by the Army or Navy Air Materiels Command. Technical expertise came from such unusual sources as women's undergarment manufacturers, infant rubber products, automobile and airplane tire companies, and leather and textile manu-

facturers. Together, these companies built a series of experimental suits in the 1940s.

The XH-5, built by Russell Colley's team at Goodrich between 1942 and 1944, looked surprisingly like the space suit we came to know in the 1960s, with big bellows resembling segments of a tomato worm's body. In his wife's garden, Colley had observed a tomato worm turn 90 degrees without any perceptible increase in pressure anywhere on its body. He and his team adapted the tomato worm's convolutions to the arms and legs of their pressure suit. Segmented bellows became the first big breakthrough in pressure suit construction. They provided the early rudiments of mobility, allowing the pilot to assume a seated position for the first time. Although "ingeniously constructed," the suit still became rigid when inflated and movement required extreme effort. The XH-5 was the most promising model but none of the suits in Table 1.1 met all the specifications laid down by the Army Air Corps. For the most part, they were stiff and still awkward and immobilized the wearer.[14]

These first full pressure suits designed for the U.S. Army Air Corps in the early 1940s protected the human body to an altitude of 30,000 m (100,000 ft.). They supplied pressure of approximately 24.1 kPa (3.5 psi) above the prevailing atmosphere, were tight-fitting, and were constructed to completely envelop the pilot in strong inelastic cloth. They had the disadvantage, however, of becoming balloonlike when inflated. The rubberized envelope encapsulating the pilot became stiff and uncomfortably warm. This made arm and leg movement almost impossible and created a need for constant ventilation to maintain proper body temperature. Ventilated air came from the aircraft's air-

Table 1.1
Early Pressure Suits

Manufacturer	Year	Model	Cost*
B. F. Goodrich Co.	1940	1	$6,100
	1941	3 (Fig. 1.5)	6,500
	1942	2, 5, 6	@3,450
	1942	6A, 6B, 8, 8A	@1,750
	1942–1943	XH-1 (Fig. 1.6), XH-1C, XH-5 (Fig. 1.7), XH-6	@3,350
Bell Aircraft Corp.	1942, 1943	BABM-9 (Fig. 1.8)**	7,051
Goodyear	1942	XH-3, XH-3A, XH-3B, XH-9, XH-9A	@ 750 2,500
U.S. Rubber Co.	1940	1 (type A)	500
		G	142
		1A, 1G, 2	2,176
	1942	XH-2, XH-4	
	1942	XH-4, XH-4A	@2,500
National Carbon Co.	1942	XH-7	

*Cost is approximate and, when known, came from Helen W. Schulz's "Case History of Pressure Suits," May 1951, Wright-Patterson Air Force Base, Attachment pp. 1–4, found in NASM archives. Also see C. G. Sweeting's *Combat Flying Equipment,* pp. 164–77. Each suit incorporated an improvement over earlier suits in the same series.

**Sweeting says BABM may stand for Dr. Walter M. Boothby, Professor John D. Akerman, and Dr. Arthur H. Bulbulian, the three persons most involved in the design of this suit.

conditioning system. It was distributed by the pressure suit ventilation garment to the pilot's extremities. Ventilated air carried the pilot's warm exhausted air and moisture to the outside of the suit. The helmets were fabricated as spherical headpieces of transparent plastic, covering additional cloth helmets and the oxygen masks the pilots wore.

Major J. G. Kearby, enclosed in an XH-3A full pressure suit, successfully tested it on October 2, 1942, to the equivalent of 19,000 m (63,000 ft.) in Wright Field's Aero Medical Laboratory altitude chamber, where a whole range of early suits in the MX-117 program had been tested. The XH series of full pressure suits

had been initially intended to help develop a usable suit for General Doolittle's crew during its raid on Tokyo in World War II. When the mission was changed from high- to low-altitude flights, pressure suits were no longer necessary.

But the crushing blow to full pressure suit development at this stage was that each suit under full pressure during flight testing reduced (sometimes to a mere ten percent) the wearer's ability to perform normal duties necessary during a military mission. For example, it was impossible to sight or operate the Norden bombsight or camera properly wearing pressurized suits. The aircraft would have to be redesigned if any of these suits were adopted

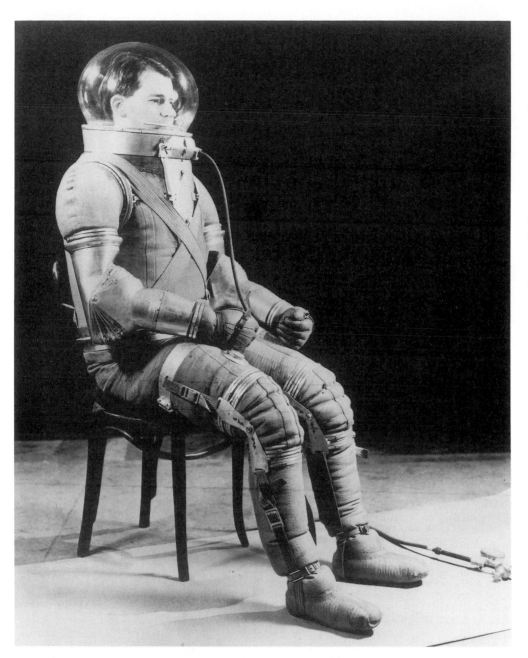

1.5
Type 3 (1941), an experimental high-altitude "Strata suit," was manufactured by the B. F. Goodrich team. All metal parts were steel and the suit was pressurized with pure oxygen. The bubble dome was detachable. The elbow joints were fashioned after a medieval suit of armor. (In *Combat Flying Clothing*, Sweeting says model 1-E may have been Type 3's factory designation.) (SI photo 84-10722)

1.6

Carroll Krupp (*left*) and Robert A. Brown (*right*), Goodrich test engineers, model the XH-1 experimental full pressure suit (1942). Goodrich received a rush order for nine suits. The size difference in these suits was important. Some suits had to be small enough for tail gunners. The engineers wanted a photo for comparison of size. They decided to pressurize the suits. Unfortunately, Brown's helmet had been improperly fixed to the back of the suit neckring. The pressure built up within his suit causing pain "like spikes" going through his ears. Someone finally noticed his distress and started pulling tubes and wiring, anything to stop the pressure build-up. Because the headpiece rested on the back of Brown's neck, he was unable to release the trigger with his own chin. In this photo, Brown's suit had been depressurized while Krupp's was fully pressurized. (photo courtesy of Robert A. Brown, SI photo 86-5007)

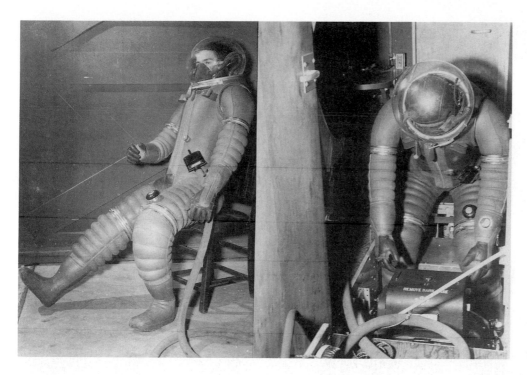

1.7

The XH-5 pressure suit, constructed by the Russell Colley team at B. F. Goodrich, was known as the tomato-worm suit (1943). Its construction was inspired by the segmented body of the tomato hornworm *(Manduca Quin-quemaculata)*, the convolutions in both suit and worm being aids to mobility. The arms and legs could be detached from the suit, making this the first model easily adapted for different wearers. Some of its technical innovations were a full-length self-sealing zipper, a detachable neckring wide enough to pass over the head, and ball-bearing shoulder joints for upper arm rotation. The rubberized fabric formed accordion-like bellows on flexor and extensor arm and leg surfaces to provide greater mobility. The suit unfortunately became rigid when pressurized. In these photos, the wearer was sitting in the photographer's position in a B-17, but found it impossible to bend forward enough to see through the viewfinder or reach the camera controls. (SI photo 87-1504)

for use. Despite ten years of research in England, Germany, France, Russia, and the United States, as well as the added stimulus of war needs, all models of pressure suits restricted mobility. Because flight test results were so disappointing, the Air Force decided that the full pressure suit design was too unwieldy and dropped it from consideration in October 1943.[15]

At first, the military did not recognize the critical need for cooperative research among aeromedical specialists, designers of flight clothing, and technicians and engineers charged with providing radio and other equipment, as well as oxygen, for the pilots. Although a skeleton team of specialists was working together by 1935, it was not until 1944 that the exigencies of war forced together a full complement of aeromedical and flight materiel specialists.

MYSTERIOUS G-FORCES

As early as 1938, General Henry (Hap) Arnold, then Chief of the Army Air Corps, sought advice from aerodynamicists to determine what the Air Corps needed to make major advances in flight. He was told to build a wind tunnel large enough to test full-scale experimental planes. And American aircraft manufacturers strove to produce planes that flew near the speed of sound. Development of advanced piston-engine fighters, new turbo-powered P-59s and P-80s, JATO (Jet Assisted Take-off) planes, and proposed futuristic X-1 and X-2 American rocket planes, as well as hints of actual German rocket-powered flights, boggled aeromedical imaginations.

Yet, there were other problems to solve. Entire B-24 and B-25 crews in mass

1.8
Experimental full pressure suits (type BABM) were originally built for Bell Aircraft Corporation as essential equipment for high-altitude operation of a proposed turbo supercharged P-39D airplane. The suit shown here was at Boeing for tests in the early 1940s. A patent for this pressure suit was filed in March 1943 by John D. Akerman, a professor at the University of Minnesota who was possibly affiliated with the Strato Equipment Company, Minneapolis, Minnesota. The suit later failed pressure tests at 10.3 kPa (1.5 psi) because of the lack of flexibility of the inflated arms of the suit, which would have prevented proper manipulation of airplane controls. (SI photo 87-6760)

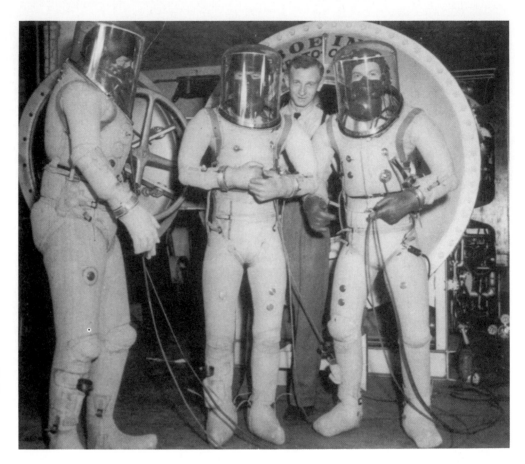

bombing raids mysteriously crashed with no evidence of any attempt to escape. It seemed possible that the spinning motion of a crashing plane pinned men down and prevented them from acting. A new term, G-force, was introduced to quantify the effect of acceleration. Researchers also feared that even if the flyers had attempted to parachute at predicted high velocities, they would have been much like a cigarette ash flicked from a speeding car. Researchers wondered if parachute-opening shock would be less at the decreased density of high altitudes.

Flight surgeon Col. W. Randolph Lovelace decided to try such a jump. It was an almost tragic skirt with death. He bailed out at 12,253 m (40,000 ft.). The temperature was −46°C (−50°F). First there was a cold blast of onrushing air. But it was the parachute-opening shock of 40 g's

● ●

Space Suits in Early Science Fiction

We are so accustomed to both the inventive imagination and scientific predictions of the big screen, that it is perhaps a shock to be reminded that writers have been speculating about space travel for more than 150 years. The nineteenth century heralded the beginnings of technological revolution and industrialization of urban life. Important novelists emerged during this developmental period whose writings made a profound impact upon generations of youthful dreamers.

In *The Unparalleled Adventure of One Hans Pfaal* (1835), Edgar Allan Poe sent Mr. Pfaal to the moon in a balloon. Poe described in detail the disastrous effects of high-altitude flying on man. The explorer suffered incredibly before he thought to enclose his car in an airtight envelope with its own oxygen supply. In *De la Terre à La Lune* (*From the Earth to the Moon*, 1865), Jules Verne also proposed encapsulating his space travelers. The proper atmosphere for breathing was maintained in their craft, and the air was purified by a chemical process. Verne also described space suits like those developed in the 1970s; he can be regarded as the father of scientifically accurate space adventure tales. Many modern scientists cite Verne as a major influence on their early dreams of spaceflight. These include Hermann Oberth, father of German astronautics who helped develop the German V-2 rocket and Konstantin Tsiolkovsky, the Russian father of cosmonautics. Wernher von Braun credited German author Kurd Lasswitz's novel *Auf Zwei Planeten* (*Two Planets*, 1897) for his own inspiration. Similarly, Robert Goddard, America's father of astronautics, credits H. G. Wells's *War of the Worlds* (1898).

In 1889, French authors Georges Le Faure and H. de Graffigny provided the space explorers in the highly illustrated *Aventures Extraordinares d'un Savant Russe* with some protective clothing and helmets resembling those used by deep-sea divers (Figure 1.9). Readers of *The New York Evening Journal* in 1898 were treated to "Edison's Conquest of Mars," an illustrated story by Garrett Serviss. His early explorers of the red planet were equipped with flight suits and helmets.

By the 1920s, popular magazines such as *Science and Invention*, edited by Hugo Gernsback, who coined the term *science fiction*, regularly featured illustrated stories about men and women who wore sophisticated suits that protected them from the harsh environment of space. Gernsback went on to publish the monthly *Science Wonder Stories* in 1929, in which Austrian Army Captain Potocnik (alias Herman Noordung, devotee of Hermann Oberth) predicted man-made earth-orbiting satellites. He also went on to describe the donning of a space suit and the need for a hand-held maneuvering unit[17] (Figure 1.10).

Cartoonists joined the ranks of space suit designers in 1929 when Buck Rogers first appeared as a syndicated comic strip character, wearing a flexible metal suit and a helmet made of metal and glass. He also carried oxygen tanks and radio phones. In 1934, *Toy World* magazine offered its readers a Buck Rogers outfit, complete with official helmet, holster, and rocket gun. Scientists and engineers later dubbed the early space programs as Buck Rogers projects.

● ●

that knocked him unconscious. Fortunately, he carried a small bail-out bottle of oxygen that continued to function supplying him with oxygen until he reached a lower altitude. The thick outer gloves had been ripped off both hands and the thin inner glove was also gone from his left hand. Lovelace was very lucky to walk away from his experiment with one frozen hand as his only serious injury. No one really knew a precise human tolerance for wind blast or gravity forces, let alone the point at which either became fatal. To study these special problems, a new component of the Army Air Force's Personal Equipment Laboratory, part of the Engineering Division of the Materiel Command, was formed. The Aero Medical Laboratory at Wright Field, Ohio, carried out research in these areas and provided support to those who developed flight clothing.

During the golden years of flight, the 1920s through the early 1940s, the National Advisory Committee for Aeronautics (NACA) brought America worldwide leadership by coordinating the talents of the nation's aeronautical science and aviation technology institutions. Army and Navy aeromedical labs joined top civilian industries to solve the problems of sending humans into higher altitudes and eventually space. Complicating the whole process in tortuous ways, however, was the constant change in specifications. Clothing and accessories had to keep pace with the evolving designs and missions of the machines the pilots were to fly.[16]

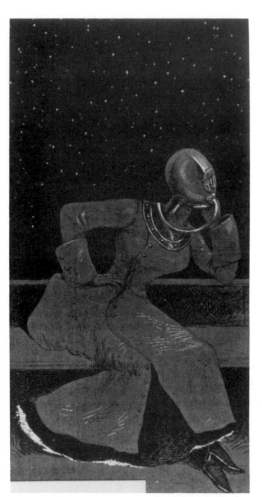

CHAPTER 2

Development of the Pressure Suit

When the Second World War broke out, the United States was quite unprepared in terms of men and materiel for open hostilities. Once the shock of Pearl Harbor wore off, Americans from all walks of life swelled the ranks of the military. Air power was built up to support the land forces. General George C. Marshall, then U.S. Army Chief of General Staff, frequently utilized air transport to ship men and materiel. By 1943, increasing altitude and duration of these long-range flights combined to emphasize the urgency of developing functional pressure suits.

Military needs continued to dictate technical progress. Dropping atomic bombs on Hiroshima and Nagasaki in 1945 was said not only to have ended World War II, but also, some thought, to have begun the Cold War. The United States felt a strong commitment to respond to communist offensives. President Dwight D. Eisenhower followed a liberal progression from President Truman's policy of reliance on nuclear strength. He felt that maintaining strategic superiority and rhetoric that indicated a willingness to use it would prove to be the strongest deterrents in the long run. Thus, the need for flight surveillance and, consequently, functional pressure suits increased.

● ●

Specialized Research, Animals, and Alice Chatham

As early as 1933, Lieutenant Commander J. R. Poppen of the U.S. Navy had exposed dogs in upright positions to pullouts four and one half times the normal force of gravity. Measurements at the carotid artery showed a decrease of pressure at the level of the head. Poppen also experimented with an abdominal belt, which, when inflated, offset the effects of greater gravity loads experienced in reentry. Poppen's early tests provided the basis for the combination of today's anti-g garments. By the late 1940s dogs were also used to develop protective clothing and equipment for aerospace programs. During the 1950s and early 1960s, the Holloman Aeromedical Laboratory trained chimpanzees, required for flight experiments demanding animals of higher intelligence, for the first demonstration flights of the Mercury program (Figure 2.1).

Alice King Chatham was a fine arts graduate from the Dayton Art Institute. Her skill in sculpting animals for outside gardens attracted the attention of personnel at the Air Research and Development Command at Wright-Patterson Air Force Base. She was persuaded to join the staff as a personal equipment design engineer/scientist in advanced biotechnology at the Wright-Patterson Aerospace Medical Lab, and developed the suit and oxygen mask for a 63.5-kg (140-lb.) Saint Bernard (Figure 2.2). The dog was used to simulate the weight of a man in drop testing of early automatic opening devices for parachutes. She also designed suits and helmets for rhesus monkeys for tests to determine the effects of wind blast on pilots ejected from aircraft at high speed. Her original restraining harness and mask for the first monkey in space was made in 1953. The little monkey rode the Aerobee Rocket in the desert where he was recovered.

Under the direction of Dr. James Paget Henry, Chatham also designed and hand made the helmet Chuck Yeager wore in October 1947 when he broke the sound barrier (Figure 2.3). A feeding device through which liquid food could be fed to someone wearing a pressurized helmet was another of Chatham's designs. A similar type of feeding device was later used for the astronauts. Chatham developed a program for making masks seal on the hard-to-fit faces of men. In 1954, bombers were flown to Patterson AFB for her to redesign and fabricate new eyeguards for bombsights and radarscopes to accommodate the K-1 pressure helmet. In 1961, she had been given the task of casting the heads of the original seven Mercury astronauts and their flight surgeon. Mold liners were made from these casts for better fitting pressure suit helmets. Chatham worked on designs for space beds, constant wear garments, and cooling garments to be used for future walks in space.

● ●

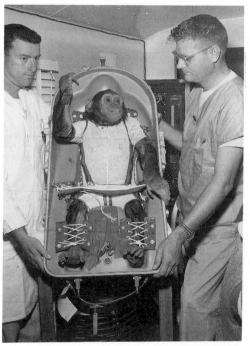

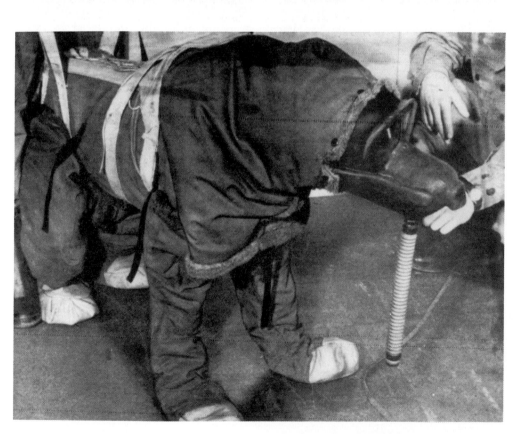

2.1 (*Top left*)
In the 1940s and 1950s, Alice King Chatham designed suits and helmets for the monkeys and chimpanzees that flew in space. Shown here is one of two chimpanzees specially trained to take part in the Mercury program. (NASA photo 61-MR2-1)

2.2 (*Bottom*)
Sculptor Alice King Chatham designed a complete suit and oxygen mask for this 63.5-kg (140-lb.) Saint Bernard. (1957, SI photo 87-15057)

2.3 (*Top right*)
This S-2 partial pressure suit evolved from Dr. James P. Henry's original S-1 suit. The full-head sealed helmet eliminated the need for an oxygen mask. The suit also eliminated the g-bladder which made it lighter than the S-1. Chuck Yeager wore this type of suit when he broke the sound barrier on October 14, 1947. Yeager wrote in his autobiography that he expected something at least like a bump on the road at that moment, but instead the unknown became more like a poke through Jell-o. Alice King Chatham, the principal designer, developed and hand-sewed the helmet for Yeager's suit. (SI photo 90-9088)

Limited mobility and a lack of automation had been major problems with earlier suits. Controls for fast pressurization of a suit that worked without human intervention were desirable in case of sudden loss of cabin pressure or other emergencies.

PARTIAL PRESSURE SUITS

Studies of pilot requirements and pressure suit problems were by then preoccupying Dr. James Paget Henry at the University of Southern California. He was aware of the problems encountered with a fully pressurized suit. He studied an alternative: pressurizing only certain parts of the suit. Thus the partial pressure suit was born. Under the auspices of the Office of Scientific Research and Development, the university built a centrifuge. Using the centrifuge and an altitude chamber, Henry began to experiment on dogs, goats, and finally humans, as he developed a partial pressure suit based on a bladder-type anti-gravity construction. Anti-g suits, commonly called g-suits, are close-fitting garments with rubber bladders, which can be inflated with either liquid or gas. Inflating them provides pressure to calves, thighs, and the abdomen to offset increased pressure of acceleration (pressure of g's experienced) to arterial blood in those areas. Blood has a tendency to pool in the lower body at high altitudes. If it is not forced back to the arterial side of the heart and recirculated to the head, fatigue, loss of vision, and unconsciousness set in.[1]

Researchers developed and tested g-suits and concepts at several locations, including the Mayo Clinic, Wright Field, the Naval Air Crew Equipment Laboratory (NACEL), and various private industrial firms. One of these firms was the David Clark Company, Inc. of Worcester, Massachusetts, which worked with Mayo specifications to develop many different models of g-suits.

An emergency partial pressure suit used the same counter-pressure applications of air-filled bladders as the g-suit to prevent undue shifts of body fluids. The suit basically consisted of an inflatable vest worn underneath a pair of close-fitting zip-on coveralls and was fed directly from the oxygen mask.

Dr. Henry wed external tubes with attached interlocking tapes to the suit fabric to pressurize the arms and legs of his partial pressure suit evenly. This system was named for the capstans, or inflatable tubes, used to pull the suit tightly against the wearer, providing mechanical counter-pressure along the limbs and sides of the body. The suit was air-cooled and was no more uncomfortable uninflated than well-tailored underclothes. It was designed to be worn under regular flying clothing and only inflated in emergencies. A full-face mask completed the enclosure for oxygen breathing. Pressure breathing was possible at extreme altitudes through the use of a properly engineered helmet, efficiently sealed to the suit. Pressure breathing is breathing in reverse. Instead of sucking in oxygen as one normally does at lower altitudes, pressure from an oxygen tank forces air into the pilot's lungs. Without a pressure suit, the pilot must physically force the air out himself, or he passes out. But once the pilot is safely suited up, the pressure suit does the work. The prototype partial pressure garment from the University of Southern California was designated as type S-1 in 1945[2] (Figure 2.4).

In 1945, the Air Force issued its request for a suit that would enable pilots to fly above the 12,000-m (40,000-ft.) limit allowed by partial pressure suits. The military also sought tactical aircraft capable of reaching speeds of 1,100 km/hr. (700 mph), the edge of the speed of sound, or Mach 1.

Near Mach 1, planes of traditional design either shook to pieces or flew completely out of control and crashed. This effect was called compressibility. Noted aerodynamicist Theodore von Kármán described an object proceeding at or near the speed of sound as generating pressure waves that bunch up in front of the object to form a condition known as shock. This creates the sudden increase in resistance often described as a sonic wall. Investigation into the effects of such high speeds and altitudes was urgently needed, and the military elected to conduct tests with experimental full-scale aircraft—the Bell X-1 and the Douglas D-558. Full-scale testing and research were also mandated to stimulate the further development of the full pressure suit (as opposed to the partial pressure suit).

The Army also experimented with altitude pressure bags (Figure 2.5) connected to a small air compressor that would serve as another rescue system in an emergency. A crew member who developed the bends could be placed inside the bag which would simulate altitude pressure low enough for relief. The bag sealed with an airtight zipper and was supplied with gauges, valves, and communication lines. Confidence in the practicality of the bags was demonstrated by the recommendation to order at least a test run supply of them.

2.4
Dr. James Paget Henry made the first partial pressure suit (S-1) by hand at the University of Southern California in 1945. (SI photo 73-10591)

2.5

The B-29 was the first combat aircraft to have a pressurized cabin where proper atmospheric oxygen and pressure could be maintained for human survival beyond 10,000 m (34,000 ft.). It was quickly discovered that the protective effects of a pressurized cabin were nullified by mechanical failure or by cabin puncture from enemy action. This pressure bag was considered an emergency rescue or "get-the-pilot-down-quick" system. Crew members zipped themselves into the bag and pressurized themselves. This type of pressure bag saved several lives as well as expensive planes. (1944, SI photo 86-10714)

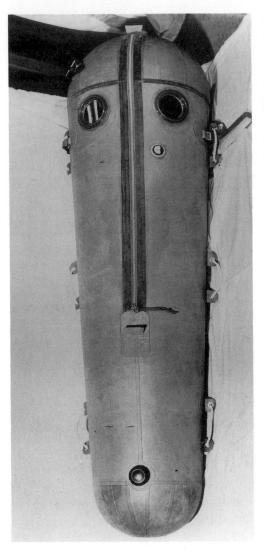

Lt. William Randolph Lovelace, Chief of the Aero Medical Laboratory, continued to direct military efforts to develop partial pressure suits. The partial pressure suit could keep a pilot alive in the case of high-altitude cabin-pressurization failure until the plane descended to a lower altitude. The S-1 partial pressure suit developed at the University of Southern California was selected on three separate occasions as the model with the most promise for pilots of the X-1 experimental aircraft.

Since the David Clark Company had long been involved in development of the g-suit, Dr. Henry called on David Clark to adapt the laboratory-model partial pressure suit type S-1 for this specific role. Dr. Henry's model left the pilot's neck partly unprotected during tests, and the skin at the back of the neck protruded alarmingly. The David Clark Company, working with Alice White and Alice King Chatham, considerably revised the S-1 in 1947 incorporating a neck seal on the helmet. Chuck Yeager, Robert Hoover, Frank K. Everest, Scott Crossfield, and William Bridgeman were among those outfitted by the David Clark Company in the commercial model of the S-1, which evolved with anti-g capability into the T-1 partial pressure suit.[3]

In 1947, the Department of Defense announced that the T-1 partial pressure suit represented the latest achievement of the Air Research and Development Command's Wright Air Development Center. Pilots who wore the T-1 partial pressure suit (Figure 2.6) were equipped with a capstan system, an anti-g feature, crash helmet, oxygen mask, earphones, microphone, goggles, defroster, and an oxygen bailout bottle. Chuck Yeager wore this suit when he flew the Bell X-1 at speeds in excess of Mach 1. Frank Everest also donned this model when he attempted to achieve higher altitudes in the X-1 in 1949.

Everest made the first operational emergency use of the T-1 suit when his plane canopy developed a crack and split as he reached supersonic speed. As designed, the T-1 suit immediately inflated

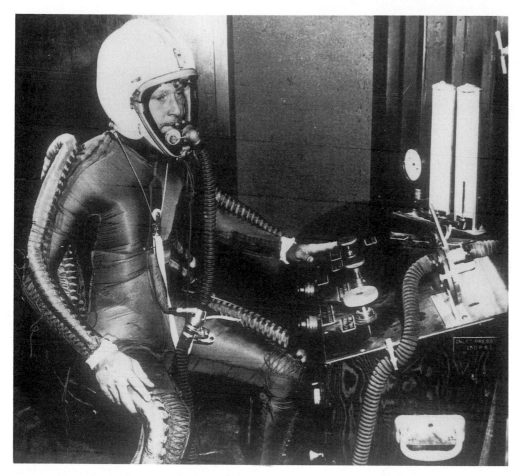

2.6
The T-1 was an improvement over the S-1, with an inflatable chest bladder in the suit and an adjusted size of capstan tubing for better comfort and fit. The suit was worn uninflated and inflated automatically whenever pressure in the airplane cabin was lost. The capstan tubes immediately tightened the cloth of the suit to withstand the internal reaction to the great pressure of oxygen being forced into the lungs. The flyer had only to exhale, for inhaling was automatic as oxygen was continuously fed into the lungs. A new two-piece K-1 helmet design (like a clam-shell) gave better visibility than any previous model. (late 1940s, 1950, SI photo 73-10579)

providing oxygen and limited protection until Everest could get his plane below 6,000 m (20,000 ft.) where he could breathe more or less normally. The suit helped save his life as well as a multi-million-dollar research aircraft.[4]

Major aircraft companies as well as the military now equipped their test pilots with the T-1 suit. Manufactured in 12 different sizes, elements of the suit could be separated and recombined to produce 36 variations that would fit 92 percent of all pilots.

In 1951, the David Clark Company received a contract from the USAF to fabri-

cate additional partial pressure suits, some with extra-high-tension fabric. Others were to have a bladder over the front of the torso to balance pressure on the diaphragm and allow for chest expansion. These were labeled S-2 partial pressure suits.

The anthropometry* unit at Wright Field had been actively studying sizing problems of pressure and flight suits since the early 1940s. The unit directed the

---

*Anthropometry is the study of human body measurements especially on a comparative basis.

2.7 (Left)

The MC-1 partial pressure suit is shown with capstan tubes inflated. Major modifications included a chest and abdominal bladder and smaller capstan tubes, features that permitted long-term use in the pressurized condition. The chest-size bladder inflated when the suit was activated. As the airman inhaled, the bladder deflated, leaving room for his lungs to expand in the tight suit. When he exhaled, the bladder reinflated and put pressure on the chest and abdomen. An XMA-1 helmet is mated to the suit. (1954, SI photo 73-10577)

2.8

The XMC-3 is shown in an early stage. The suit had been used by the laboratory primarily for studying physiological stress encountered by wearers of pressure suits. The XMC-3 had capstan tubes and was fitted with a full torso bladder rather than simply a chest bladder. The suit could prolong protection at higher altitudes, but it still limited the wearer's mobility and had ventilation and comfort problems. (SI photo 73-10593)

next stage of development and the result was the MC-1 model (Figure 2.7). The first MC-1s did not solve all fitting difficulties. Because pressurized gloves were worn with the MC-1 and had not been worn with T-1 or S-2 suits, sleeve lengths created a new problem. The modified MC-1 was provided with additional sleeve-length tubing, which could be adapted to fit the individual wearer.[5]

Military requirements during the cold war and Korean conflict intensified the demand for jet-propelled interceptors. Plans for the B-52 and B-58 bomber aircraft were introduced by the Air Force in 1953. These planes, designed for altitudes between 15,000 and 21,000 m (50,000 and 70,000 ft.), sparked renewed research efforts to develop a successful long-term high-altitude suit. In March 1956, requirements for the B-36 and B-52 planes

to drop the H-bomb from the edge of the stratosphere over the Bikini Atoll necessitated development of dependable pressure suits. Suddenly the XMC-3 (Figure 2.8), an experimental partial pressure suit used by the Aero Medical Laboratory for research to develop high-altitude equipment, was put on active duty for the bomber crews. The Air Force canceled its contracts for other partial pressure suits, but soon realized it had acted pre-

maturely. The XMC-3 limited the wearer's mobility and gave users problems with ventilation and comfort. The MC-1 and T-1 suits were still in demand.

The Air Force placed an order in March 1956 for T-1A suits, which incorporated larger capstan tubing and an improved anti-g bladder. Issued with this suit was an improved MA-1 helmet that could be worn with both full and partial pressure suits. Pilots complained that the earlier K-1 and K-2 helmets impaired their vision. The MA-1 helmet later proved to have its own set of problems with weight, oxygen leaks, and an unstable sealing system. The last partial pressure suit model, the MC-4A with an MA-2 helmet contained additional anti-g protection and was made in limited numbers. It was declared the standard by January 1960.[6]

Since 1943, the Air Force and the Navy had been working under a joint agreement that limited the Air Force to development of partial pressure suits. Although the Navy had been assigned the task of developing a full pressure suit, its short-term approach and requirements could be met by the T-1 partial pressure suit. By 1953, Air Force crews required more protection than was offered by the partial pressure garment. Strategic Air Command crews could no longer afford to abort a mission simply because of loss of cabin pressure; they had to be able to remain at altitudes over 15,000 m (50,000 ft.) for considerably longer periods to avoid enemy action, reach their target, and return to home bases. For economic reasons, the two services agreed that the Navy would concentrate on full pressure suit research, while the Air Force would continue its partial pressure suit program; but both services stepped up their efforts.

FULL PRESSURE SUITS

The Navy full pressure suit developed by June 1954 under Air Force contract with B. F. Goodrich, model 2-A (Figure 2.9), was cumbersome to don and doff. The suit lacked mobility, proper ventilation, and comfort and the helmet restricted vision. Because of these failings, the Air Research and Development Command ordered the Aero Medical Laboratory to develop a suit to meet Air Force needs. The Aero Medical Laboratory invited 30

2.9

B. F. Goodrich made this model 2-A, a development model of a full pressure suit, for the Navy in 1954. The suit was a combination of soft rubberized cloth and heavy rubberized material. The full pressure helmet made an oxygen mask unnecessary, but it restricted vision; and the whole assembly lacked mobility and proper ventilation. (SI photo 73-10592)

2.10
In 1958, XMC-2-ILC was the first International Latex suit delivered to the U.S. Air Force. The limbs were constructed with numerous rubber coils and molded indentations, which helped to keep the suit's internal gas pressure constant. (photo courtesy of ILC Dover, Inc.)

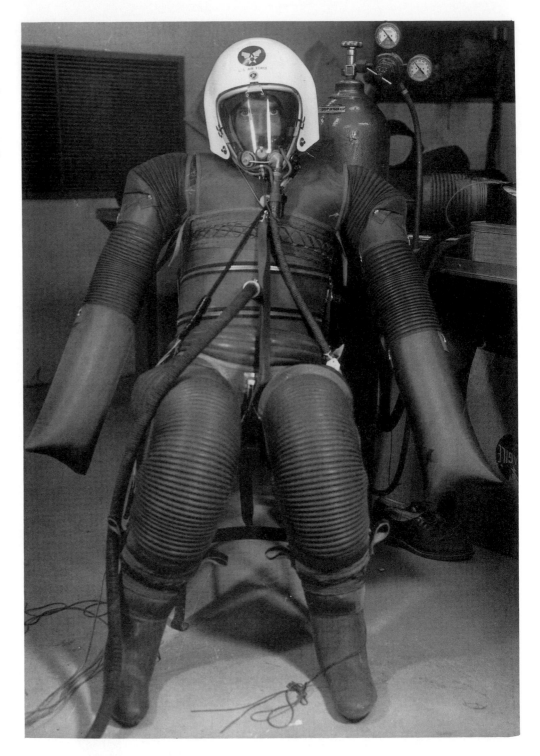

different companies to bid on new designs and several contracts were awarded during the spring of 1955. The two eventual products of these efforts were designated the XMC-2-ILC (Figure 2.10), manufactured by the International Latex Corporation, a rubber and fabric products company in Dover, Delaware, and the XMC-2-DC (Figure 2.11) from the David Clark Company.[7]

The full pressure suit program had produced only discouraging results. When pressurized, the fabric used had a tendency to balloon and stretch until almost rigid, restricting pilot mobility. Pressure seals further hampered arm and wrist movement. A monumental breakthrough came through in 1956 when the David Clark Company and the Air Research and Development Command* gave the U.S. Air Force its first practical full pressure suit.

This was the first type of full pressure suit that did not incorporate the rigid tomato-worm bellows. The USAF A/P22S-2 breakthrough incorporated a soft slip-knit *linknet* (like fishnet) nylon layer constructed to form a restraint on the inflated pressure capsule around the pilot, so that it did not balloon shapelessly, but still gave mobility and comfort.[8]

Modifications to the suit known as the Navy Mark IV made it compatible with Air Force aircraft. The resulting suit became the A/P22S-3. The Air Force, meanwhile, studied a quick-donning pressure suit with more bulk and insulation, the CSU-4/P. It was the improved

*The Air Research and Development Command is now designated the Air Force Systems Command (AFSC).

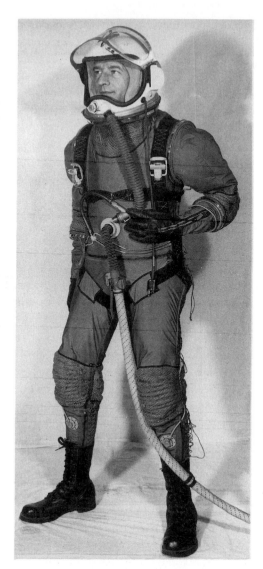

2.11
The XMC-2-DC was the David Clark Company's "assembled from components" suit delivered in 1956. The XMC-2-DC went through various development stages until it became known as the A/P22S-2. (SI photo 73-12887)

A/P22S-2, however, that became the standard Air Force full pressure suit. It provided the best yet combination of altitude and exposure protection. It permitted flight to any altitude and was comfortable if properly vented, whether pressurized or not. Its helmet defogging capability was better than any to date. This suit is considered the precursor to the Gemini suit (Figures 2.12 and 2.13).

2.12 (Right) **and 2.13**
(Facing page)
The A/P22S-2 from the
David Clark Company
provided the best combi-
nation of altitude and exposure protection then
available. The suit assembly was covered with an
aluminized layer to reflect
heat and ultraviolet rays
and to offer protection in
case of fire. It was worn
by Neil Armstrong in
1961, when he made
seven successful flights of
the sophisticated X-15
research aircraft. (SI photos 87-6759 and 88-27)

Pilots testing the new X-15 research aircraft for NACA* and the Air Force were outfitted with this new full pressure suit. The X-15 was designed to fly at speeds of Mach 4, 5, and 6 and to altitudes exceeding 90,000 m (300,000 ft.), bridging the gap between atmospheric flight and spaceflight. Craft and pilot were exposed to high-gravity loads and the exterior surface of the plane was exposed to temperatures in excess of 650°C (1,200°F). Test pilot Scott Crossfield wore an A/P22S-2 full pressure suit on September 12, 1959, in the X-15's first rocket-powered flight. Crossfield had been working on pressure suit development with David Clark since 1951. He said some parts of the first suit they designed were fabricated on Mrs. Clark's sewing machine.

Twelve pilots successfully flew the X-15 between 1959 and 1968. A maximum speed of Mach 6.7 and a maximum altitude of 107,960 m (354,199 ft.) were attained during this period. Eight pilots achieved astronaut rating by flying to altitudes beyond 80 km (50 mi.). The X-15 flight research was an enormously important and successful project to both NASA and the military. Innovative technology and materials came from the X-15 program such as improved use of Russell Colley's tomato-worm convolutions at suit joints, the use of high-technology materials for flight clothing topped by the link-net restraint, and semi-automatic pneumatic controls. The X-15 program implemented reliable, maneuverable, and

MODIFIED DAVID CLARK FPS

comfortable space suits with constant volume and constant pressure.

The X-20 program, known as Dyna-Soar for "dynamic soaring," was an idea ahead of its time. It was anticipated that the X-20 would function both as a rocket-boosted orbital spacecraft and as an airplane for landing purposes. The X-20 suit was derived from the X-15 suit and eventually evolved into the Gemini suit (Figure 2.14). Because the Air Force decided to consolidate its efforts into plans for the Manned Orbiting Laboratory (MOL), the Dyna-Soar program was canceled in 1963. Its research, however, was of value to the military and to NASA.

The Air Force's A/P22S-2 suit still lacked sufficient mobility, as well as a fully automatic pressurization system, and had considerable leakage. It de-

_____

*To address the growing potential for spaceflight, NACA evolved into the National Aeronautics and Space Administration (NASA) on October 1, 1958. Military matters formerly handled by NACA would now be handled by the Department of Defense.

2.14

The 1962 X-20 pressure suit was intended for use in the Dyna-Soar aircraft. It was similar to the A/P22S-2 in structure, with an aluminized cover layer, and included boots and gloves. The bubble helmet was supported by the internal suit structure rather than separately joined and offered good visibility. (SI photo 83-15508)

pended upon the manual regulation of suit-mounted pressurization components. Meanwhile, B. F. Goodrich had reentered the picture. Russell Colley, who had perfected Wiley Post's successful pressure garment and engineered the experimental XH-5 tomato-worm suit, headed a group of dedicated engineers who developed the

"Mark" series of full pressure suits in conjunction with the U.S. Navy Bureau of Aeronautics.

These four models of full pressure, pneumatic suits, Marks I, II, III, and IV, shared similar characteristics. They featured a close-fitting helmet with a sealed visor, separated from the rest of the torso by either a neck or face seal. Dry oxygen flowing over the interior visor surface and a chemical coating were used to prevent fogging. Suit entrances were sealed by pressure-sealing slide fasteners. A cable arrangement attached to the helmet neck-ring and torso helped control helmet rise and suit elongation during pressurization. The interior layer of the suit consisted of a neoprene-coated nylon and Helenca stretch-knit fabric. The outer layer, or restraint, made of non-stretch nylon twill, controlled the shape and ballooning of the interior gas-retaining fabric. A ducting system distributed ventilation gas.[9]

The biggest breakthrough in the Navy/ B. F. Goodrich full pressure suit program occurred in 1952. Russell Colley and his group of engineers solved most of the mobility problems with the invention of a swivel joint of airtight rotating bearings and fluted joints. The 1952 Mark I Model 1 suit reflected Colley's original tomato-worm suit with segments resembling tiny rubber tires. This suit had been rejected earlier because it was cumbersome. The Navy, meanwhile, solved another critical problem. The service designed an aneroid controller that automatically maintained suit pressure at 24.1 kPa (3.5 psi) inside the suit no matter what the outside pressure became. An aneroid barometer measures air pressure by the action of air within a vacuum, not by height of a fluid column. Earlier suits were dependent upon manual regulation of pressure.[10]

The first improved suit to emerge from B. F. Goodrich's contract of 1952 with the Navy was designated the H model. Colley and his team adopted Clark's fabrication methods by vulcanizing their material: The rubber tomato-worm segments were treated chemically to give them elasticity, strength, and stability. They were now like small semirigid accordion pleats that gave the suit considerable mobility. This full pressure suit incorporated a retractable visor, a form-fitting headpiece integrating a crash protecton shell, an oxygen-defogging feature, a breathing regulator mounted on the face mask, and an opening for the nose and mouth. An L model incorporated a detachable headpiece for easier donning. With the production of the M suit in 1954 the company met another milestone in successfully miniaturizing the oxygen equipment. Before this, suits utilized an oral-nasal breathing mask that limited visibility, felt uncomfortable, and was complex in construction. In the M model a face seal replaced the face mask. This suit also included detachable gloves.

Another identifiable developmental stage of the B. F. Goodrich suit was the R model, designed in 1953. It featured a sit-stand-sit capability, a tinted visor, a zippered sizing band adjustment feature, wire palm restraints for the gloves, and a belt-type entrance zipper. The S model suit, developed to be watertight, had detachable boots, improved helmet adjustments, and a unique, new tiedown system to prevent the helmet from rising under pressure. It was the S model suit that was manufactured in limited quantities and designated the Mark I, Model 3 (Figure 2.15).[11] The Mark II model was a soft suit of lightweight rubberized fabric and was probably the most flexible (increasing

mobility) of all the Mark series models. Its helmet had much improved visibility as well as more comfort than previous naval headgear[12] (Figure 2.16).

The Navy was concerned about the overall bulk and weight of the Mark II and found that the shoulder bearings and helmet tiedown cables still caused discomfort. The Navy opted to trade ease of mobility for lighter weight. The Arrowhead Rubber Products Company (a subsidiary of Federal-Mogul Corporation) and B. F. Goodrich were rival bidders; each presented a design for the Mark III suit. Both were lightweight models with

2.15

The Mark I full pressure suit, manufactured in 1956 by the B. F. Goodrich Company, had been rejected in its development stages because it was cumbersome. But the Navy solved an important problem with the Mark I by developing an aneroid controller that automatically maintained suit pressure at 3.5 psi, no matter what the external pressure was. (SI photo 84-10710)

for another use in the 1960s: NASA's Mercury astronauts would depend on it for protection in a new arena—orbital spaceflight.

The steady progress in flight clothing development could not keep pace with advancing technology. In the aftermath of World War II, the German scientists who created the V-2 rocket missiles split up, taking their knowledge to two competing nations: Wernher von Braun and his team emigrated to the United States; Helmut Gottrup's team was recruited by the Soviet Union. On October 4, 1957, the Soviets launched *Sputnik 1,* the first unmanned spacecraft to reach beyond the earth's atmosphere. The demand for space clothing quickly followed.

Table 2.1
Partial Pressure Suits

Manufacturer	Year	Model	Special feature	Cost*	Source
USC	1943	S-1	Capstan tubing	$2,500	Mallan
D. Clark Co.	1947	T-1	Capstan, anti-g	1,500	Schulz
D. Clark Co.	1951	S-2	Capstan, bladder	3,300	Schulz
D. Clark Co.	1954	MC-1	First suits sized		
D. Clark Co.	1956	MC-3	Firewel regulator	769	AFCSG-11
D. Clark Co.	1956	MC-4A	Extra anti-g		
D. Clark Co.	1956	T-1A	Larger capstan		

*Cost is approximate, where known. David Clark once commented that he almost had to be King Solomon to determine exact costs of suits because there were so many developmental changes involved, as well as other contractors.

Table 2.2
Full Pressure Suits

Manufacturer	Year	Model	Special feature	Cost	Source
ILC	1956	XMC-2-ILC			AF
D. Clark Co.	1956	XMC-2-DC			AF
D. Clark Co.	1956	A/P22S-2	Linknet	$3,891	AFCSG-11
B. F. Goodrich Co.	1956	Mark I*	Aneroid controller		Navy
B. F. Goodrich Co.	1957	Mark II	Airtight rotating bearings		Navy
B. F. Goodrich Co.	1958	Mark III	Ventilation system		Navy
Arrowhead	1958	Mark III	Lightweight		Navy
B. F. Goodrich Co.	1959	Mark IV	Improved suit, helmet, gloves		Navy

*In 1954, the Navy awarded Goodrich a contract for $130,000 to develop this series of suits (approximately $5,000 apiece). (Mallan, p. 130)

CHAPTER 3

Mercury
Space Suits

n 1950 the Korean conflict had shaken American confidence. The United States began developing guided missiles for all-purpose interception and weapons. Rocket motors, using both solid and liquid fuels, replaced jet propulsion. The necesssary technology and first-generation boosters evolved from the military missiles programs. Both the United States and the Soviet Union pursued exploration of the upper atmosphere with their supplies of captured V-2s. In fact, this early V-2 research in America is considered the beginning of our space program. And when these supplies were exhausted, they developed newer and more reliable research rockets. Development of intermediate and long-range missiles also continued.

On November 1, 1952, the Atomic Energy Commission detonated the world's first thermonuclear explosion over the Eniwetok Atoll in the Pacific. The Soviets exploded their first thermonuclear device in 1953. These events spurred intercontinental ballistic missile (ICBM) development (second-generation boosters). At the Army Ballistic Missile Agency (ABMA) in Huntsville, Alabama, on September 21, 1956, a modified Redstone missile named *Jupiter-C* was successfully test-launched. Since President Eisenhower did not want a military missile to be the carrier of the first U.S. satellite, the fourth (satellite) stage was a dummy stage.

In December of that same year at Cape Canaveral, Florida, a group of scientists and engineers set out to launch a satellite as part of their Project Vanguard. The first stage was a Viking, with an Aerobee second stage, topped by a Vanguard third stage. It was to have been the first of three planned satellite launches announced in conjunction with the International Geophysical Year activities. Unfortunately, swarms of reporters took it to be America's response to Sputnik and descended upon the launch site. During the nationwide television broadcast, the Vanguard exploded. Secretary of Defense McElroy took a dim view of this disaster. He ordered the Army Ballistic Agency at Huntsville to revive its 1954–55 Project Orbiter proposal, the one originally overruled by the Defense Department in favor of the civilian Project Vanguard.

Wernher von Braun's group of scientists hurriedly converted their *Jupiter-C* reentry test vehicle into a satellite launcher. On January 31, 1958, the *Jupiter-C,* renamed *Juno 1,* boosted *Explorer 1* into orbit. America had its first weather satellite. In 1952, the ABMA had taken a modified V-2 called the *Hermes C-1* and modified it further for high-mobility field development. A Navaho booster engine was added and it was officially named Redstone. The second successful military rocket, the Atlas, was developed by the Air Force as the first ICBM. Before sending humans into space, both the Americans and the Soviets successfully launched animals such as mice, chimpanzees, monkeys, dogs, fruit flies, bees, and hamsters into orbit on their modified V-2s. Redstone and Atlas rockets would successfully boost the Mercury program into space.

In October 1958, President Eisenhower established the National Aeronautics and Space Administration (NASA) to manage the U.S. civilian space program, thereby dividing spaceflight goals between the Department of Defense and NASA. The military would pursue space projects of a martial nature, while NASA would manage those activities "devoted to peaceful purposes for the benefit of all mankind."[1] Included among those assignments given to the civilian agency was the U.S. manned space program named Mercury. In part, NASA was established in reaction to the early achievements in spaceflight of the USSR. The Soviets had launched the first artificial satellite, *Sputnik 1,* in October 1957 and, during the following months, continued to orbit spacecraft that were bigger and could support animal life. Many experts believed that the Soviets would soon attempt a manned mission.

A year before NASA opened its doors, the Air Research and Development Command (ARDC) directed the Wright Air Development Center to seek the quickest method of putting a human into space. As part of this search, ARDC approved $445,000 for the study of an internal ecological spacecraft system capable of sustaining a person in orbit for 24 hours.[2]

Critical to any spacecraft life support system would be a protective suit which pressurized if there was a sudden leak or decompression of the craft. In such an emergency, delicate barometric sensors would signal the pilot to close his helmet face plate, after which the suit would automatically inflate. In essence, the wearer would be enveloped in an oxygen-controlled, pressurized mini-spacecraft.[3]

The Mercury spacecraft was to be a simple, bullet-shaped single-occupant

spacecraft, designed to evaluate a person's ability to withstand the rigors of orbital flight. Several companies, including International Latex Corporation and the David Clark Company, competed with B. F. Goodrich to produce the Mercury suit (Figure 3.1). The space suits that Mercury astronauts,* as the pilots were dubbed, would wear had evolved from the Mark IV pressure suits developed by

the Navy during the 1950s and adapted for supersonic flight. NASA awarded the B. F. Goodrich Company, chief supplier of U.S. Navy full pressure suits, a contract in July 1959 to fabricate 21 pressure suits and spare parts for the Mercury program at a total cost of $98,000.

Among the first to learn about the project was the family of Russell Colley. While they sat around the dinner table, Colley took a piece of paper and drew the outline of a spacecraft that was to contain a man above an enormous rocket that would boost him into space. His family leaned over the drawing. His daughter Barbara said, "A man is going to be in that?" Her father replied, "Yes, yes. Not an airplane, but a rocket."

3.1

Three of the contenders for producing the Mercury space suit in 1959 were the David Clark Company, *(prototype shown left)*, International Latex Corporation *(middle)*, and the B. F. Goodrich Company *(right)*. Goodrich was selected to design and develop the Mercury pressure suit. (photos courtesy of Edwards Air Force Base, NASM photo 92-620)

---

*It is thought that the term *astronautics* was first used by Belgian science fiction author J. J. Rosny in 1926. Astronautics and astronauts had certainly been accepted into the language by 1930. These words combine the Greek words *astron* (star) and *nautes* (sailor); thus, an astronaut is literally a sailor among stars.[4]

3.2

The XN-3 Mercury space suit was a modified Navy Mark IV "quick fix" suit to be used for operational research. It incorporated a number of design changes from the Mark IV Model O type lightweight full pressure suit. The outer ply of nylon is aluminized for better heat resistance. The leather boots are coated white for the same reason. The helmet detaches from the suit by means of a disconnect only. No neck bearing is used on this suit. Other design changes were required because this suit was used with a closed oxygen system, thereby eliminating the need, among other things, of a headpiece face seal diaphragm. (June 17, 1959, SI photo 83-15845)

Colley said there was a great deal of secrecy wrapped around the program. One day he got an urgent message to bring a pressure suit to the Navy base. It was packed into a box a little over three feet high marked "confidential." Colley showed his special pass to get into the base. The guard wanted to examine the contents of the box. It took many phone calls to get that box inside the building without having to disclose the secret.

The suit had zippered openings for donning and doffing, a neoprene-coated nylon layer to prevent leakage, an airtight neckring bearing, fabric-fluted shoulder and knee joints to allow the astronauts to move, and an over-garment fabricated of high-temperature-resistant aluminized nylon. The vulcanized, double-walled garment with a perforated inner wall permitted the body pores to breathe. Air flowed into the inner suit through a waist connector, circulated throughout the suit, and exhausted through a pipe in the helmet. The helmet locked onto the suit's special padded neckring.[5]

The first pressure suits delivered to the agency were "quick fix" Navy Mark IV suits, designated XN-1 through XN-4 (Figure 3.2). NASA and its contractor suffered several irksome problems with these slightly modified suits during their trials, such as stretching fabric, user discomfort, and poor air circulation. Goodrich requested an additional $100,000 for more research and development. New materials with a reduced tendency to stretch were sought. Engineers investigated the possibility of allowing elbow and leg joints to assume bent positions upon pressurization. Studies were made of the compatibility of the molded spacecraft couch* with the astronaut's pressure garment.[6]

To facilitate space suit alterations, Goodrich engineers used body molds. Prospective astronauts were dressed in

---

*A custom-contoured couch gave Mercury astronauts solid support under high-g stresses such as during lift-off and reentry. Blood tends to pool in the abdomen at moments of high stress. Researchers found that if an astronaut lay on the couch with legs positioned higher than the head, blood was more evenly distributed over the length of the body.

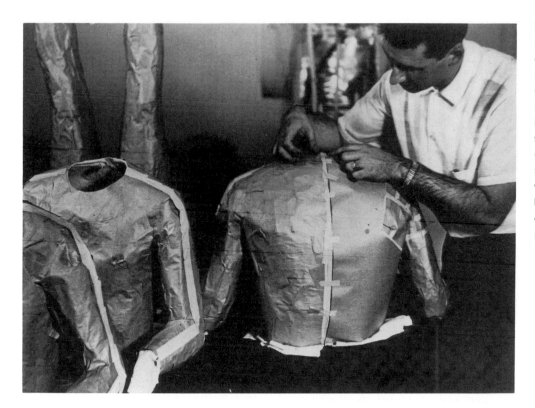

3.3
Evaluation of test suits indicated modifications for increased mobility and enhanced spacecraft compatibility. Careful sizing based on body molds such as these helped produce individually tailored space suits. Stretching which had been a problem was significantly reduced. (1961, NASA photo S-65-3528)

long underwear and covered all over with wet strips of brown paper tape. When dry, tape and underwear were cut away. The resulting mold was used to manufacture the tailored suit (Figure 3.3). Weighing about 10 kg (22 lb.), the suit was equipped with 13 zippers to assure a snug fit. Each astronaut was provided with three suits—one for training, a second for flight, and a third as backup—at an approximate cost of $5,000 each.[7]

NASA's Space Task Group,[8] the organization assigned the project management of Mercury, convened a design conference in May 1960 to further examine the Goodrich suit. Several design changes to the Mark IV resulted: Existing shoulders were replaced by segmented shoulders with diagonal pleat panels at the rear; straps were added to prevent shoulder rise

and underarm cutting during pressurization; and an extra sponge-rubber insulation layer was dropped because engineers had determined that the spacecraft would sufficiently protect the astronauts from heat build-up during reentry. The seven Mercury astronauts also chose curved-finger gloves with one straight finger to assist in pushing guarded buttons on the instrument panel. Extra threads woven into the glove material roughened the texture for easier manipulation of push buttons and toggle switches.[9]

Helmets, which had to fit perfectly, accounted for about half the suit's cost. In Los Angeles, Alice King Chatham made a cast of each astronaut's head (Figure 3.4). From these molds, helmet liners, called "Lombard" after their designer, were formed with an outer covering of leather.

3.6 *(Top)*
Virgil "Gus" Grissom is suited up before entering the Mercury spacecraft *Liberty Bell 7*. (July 21, 1961, NASA photo 61-MR4-76)

3.7
John Glenn's gloves incorporated tiny lights in the fingertips which helped him read manuals and operate the spacecraft during that part of his orbit on the dark side of the earth. (February 20, 1962, NASA photo 62-MA6-183)

lowed a low-residue diet for several days before flight to help stave off the possibility of nausea and keep bodily eliminations to a minimum.[12] During his nine-hour flight, Schirra complained about spending a good bit of time adjusting his suit's temperature settings.

Gordon Cooper orbitted the earth 22 times during the final Mercury mission, MA-9, which took place in May 1963 and lasted 34 hours. Cooper's suit for this mission incorporated boots, new shoulder construction, improved gloves, a helmet-mounted thermometer, new microphone, and a mechanical visor seal (Figure 3.8). Temperature fluctuations also plagued Cooper, but at the end of the thirteenth orbital pass he reported his surprise at being comfortable enough to nap soundly.

3.8
In May 1963, Gordon Cooper made 22 orbits of the earth wearing this MA-9 suit. Cooper's suit reflected many changes since the early Mercury flights. (NASA photo S-63-1754)

Just before reentry, he noted that short circuits had fouled the automatic stabilization and control system, which he was able to override with a manual system. Carbon dioxide levels had risen within his suit as well as within the cabin. Cooper's manual intervention quite likely saved his mission and underscored the usefulness of human backup to automatic systems.[13]

The greatest lesson learned from the Mercury flights was probably the unique importance of people to machines. The Mercury program began with a machine that had a man in it. But by the end of the program, it truly became a manned spacecraft. NASA also learned a good deal about the physiological effects of weightlessness, vital preparation for the "space walks" and long-duration two-man missions of the Gemini program.[14] (See Appendixes 4 and 5 for the chronology of Mercury suits.)

Table 3.1
Mercury Materials
Pressure Garment Assembly (PGA)

Layer	Material	Function
1	HT-1 aluminized nylon	Abrasion/flame resistance
2	Neoprene-coated nylon	Leakage prevention
3	Vulcanized, double-walled nylon	Inner wall ventilated for air circulation and cooling
4	Long-john underwear with waffle-weave patches	Comfort Cool-air circulation

Note: Extra insulative layers (such as those in Gemini and Apollo suits) were not used for Mercury because engineers determined that the spacecraft would sufficiently protect the astronaut from heat build-up during reentry.

CHAPTER 4

Gemini Space Suits

Even before NASA launched its first manned orbital Mercury mission, President John F. Kennedy had proclaimed that the United States would send a man to the moon and return him safely by the end of the 1960s. This extremely ambitious goal for the young space agency would be gained only by painstaking work. Apollo crews would indeed land on the moon. But first, Gemini crews would need to explore further the limits of their spacecraft and their own abilities to perform useful work in the weightless environment of space.

As the Mercury program faded into history, many questioned its necessity and $400-million cost, as well as the proposed costs for the Gemini and Apollo programs. Sending humans beyond earth's gravitational field required bigger and more complicated boosters than Mercury's Redstone or Atlas. Other questions arose. How would metals, plastics, sealants, insulators, lubricants, moving parts, flexible parts, surfaces, coatings, and liquids react in outer space?

A slightly modified Titan II weapons system was suggested as the booster for Gemini spacecraft in 1961. The Titan was a longer range, higher thrust, second-generation ICBM. The Air Force had already expended considerable time and effort adapting the Titan to manned spaceflight in the Dyna-Soar program intending it to become our first manned spaceflight program. The Titan was capable of lifting considerably heavier loads than the Mercury Atlas. Atlas used a cryogenic propellant system (operating at very low temperatures, i.e., liquid oxygen), while Titan used a much simpler hypergolic propulsion system (fuel and oxidizer burn on contact, no complicated ignition system is necessary) that used storable liquid propellants. Since orbital rendezvous was a major objective, the manned Gemini spacecraft would attempt docking with an Air Force Agena stage placed into orbit by an Atlas.

The estimated cost was in the neighborhood of $500 million. The actual final cost of the entire Gemini program totaled $1.29 billion: $790.4 million for spacecraft development, $417.4 million for the launch vehicles, and $82.3 million for the support system.

When Kennedy was assassinated in November 1963, matters of spaceflight passed into the hands of President Lyndon Johnson. Johnson tended to look toward the bigger picture of where the United States would be in ten years. He sought to make America's programs prove the success of our system and our way of life. Johnson was less moved by the adventure and wonder of the space program. But, he pledged to see the Gemini and Apollo programs through to their conclusions, as well as to seek out Soviet cooperation in a joint venture.

By 1964, Brezhnev replaced Khrushchev in Moscow and the Chinese had exploded their first atomic bomb. A cold war had settled over the world. LBJ saw it as a battle between those who wished to be free and those who wanted to enslave mankind through communist rule. In his Great Society, Johnson saw American space exploration as one way to fight communism, because it assured our scientific and technical superiority.

GEMINI SPACE SUIT SPECIFICATIONS AND EVOLUTION

Ten two-man crews would fly increasingly complex and lengthy missions during the mid-1960s for Project Gemini. The astronauts would need intravehicular pressure suits that provided freedom of movement and comfort for long durations. To allow the wearer to work outside the spacecraft, the Gemini suit would also have to provide protection from the environment of space itself.

In April 1962, NASA's Manned Spacecraft Center (MSC), Houston, Texas, awarded the Aerospace and Defense Products Division of the B. F. Goodrich Company a contract valued at $209,701 to design, develop, and fabricate prototype pressure suits. Other contracts went to Arrowhead Rubber Products in Los Alamitos, California, and Protection, Inc. in Gardena, California. The Air Force also contracted the Rand Development Corporation for a space suit study and a demonstration model (Figure 4.1).

When Goodrich began work on their contract, there were two identifiable NASA pressure suit programs, neither of which was associated with any manned space program in particular. The contractor's original instructions were to produce four successively improved versions of an advanced pressure suit and two prototypes of partial-wear, quick-assembly full pressure suits. NASA amended this contract in September 1962 to identify the suit project with the Gemini program. The Manned Spacecraft Center's Life Systems Division stipulated the special needs

4.1

The result of an Air Force/Rand Development Corporation space study (performed by A. S. Iberall, as the principal investigator between 1951 and 1970), this suit was an experimental prototype. Although human skin stretches during body motion, there is virtually no stretch along certain lines called "lines of nonextension." Pressure suits were developed to permit natural mobility and minimize ballooning. (1963, SI photo 87-15056)

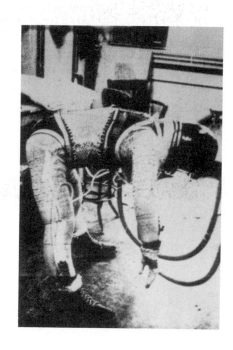

of Gemini astronauts inside and outside their spacecraft, and the contractors set to work to design a Gemini-specific suit. Model GX1G* was the first prototype (Figure 4.2).

Designs for the Gemini garment constantly changed during its development. Mission-related requirements called for 14 days' continuous wear, pressurized mobility compatible with spacecraft crewstation design, pressurized operation at 25.5 kPa (3.7 psi), and maximum unpressurized comfort and mobility. Evaluation exercises and actual flight experience led to further refinements and improvements.[1]

NASA flight crews wanted comfortable space suits. Their major requirements were reasonable: that their suits fit without stress on body pressure points, that body thermal balance be maintained efficiently by the environmental control system in conjunction with suit ventilation distribution, and that suits be equipped with a proper waste management system. Some Gemini crew members would face an unprecedented challenge—tethered extravehicular activity (EVA)—and have even more demanding tasks set by mission control. Gemini space suits would have to provide astronauts with an EVA duration capability of between 25 minutes and an hour, ventilation system hardware, and methods of maneuvering in proximity to

*All space suits accepted by NASA were given designations beginning with G for Gemini. X stood for experimental. The next letter represented the maker. Thus GX1G was the first Gemini experimental prototype from Goodrich. The final number represented the individual suit in a series: G1C-1 was the first in a series of space suits that the David Clark Company produced for the Gemini program and was used for engineering tests prior to training and qualification.

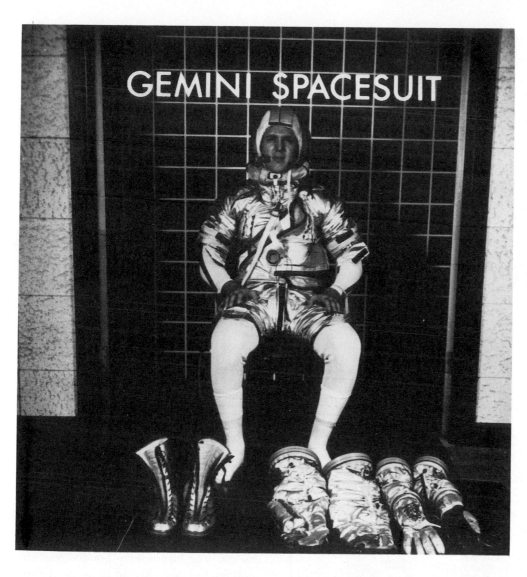

GEMINI SPACESUIT

4.2 *(Facing page)*
In 1962, B. F. Goodrich was awarded the first contract to design Gemini space suits. The GX1G was the prototype of this advanced full pressure suit. Note the flexion possible in the arms with the suit pressurized to 24.1 kPA (3.5 psi). (NASA photo 62-1976)

4.3
The G2G was the second prototype developed by Goodrich to be a partial-wear, quick-assembly full pressure suit with detachable accessories. (1963, NASA photo 72-H-735)

the spacecraft and incorporate a super-insulation cover layer worn over the pressure suit for thermal protection.

In November 1963, Goodrich delivered the second of two partial-wear suit prototypes, which included detachable sleeves, legs, and helmet (Figure 4.3). The company received an order to provide NASA with 14 more suits based on this design in various sizes. The prototype project was now designated G2G, with the suits numbered G2G-1 through G2G-15. After extensive testing and design changes, model G2G-8 was rejected in 1963 in favor of a suit designed by the David Clark Company. This new Clark suit incorporated Goodrich helmets, gloves, and hardware (Figure 4.4).

The David Clark Company's Gemini suit descended directly from the X-15/X-20 suits originally pioneered for the pilots of supersonic aircraft. Weighing in at 11 kg (24 lb.), the David Clark design solved the constant volume/constant

4.4

The GIC suit from the David Clark Company, Inc., was selected following the rejection of the G2G. This Gemini suit, delivered in early 1963, incorporated Goodrich helmets, gloves, and additional hardware and was basically used for engineering tests. (NASA photo S-62-8074)

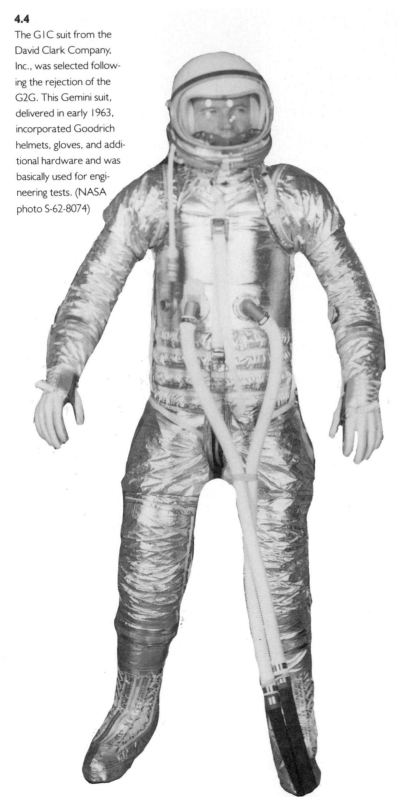

pressure problem in a new way. Mercury suits had provided constant volume and pressure with restraining cords sewn about the elbow and knee areas. These inelastic cords prevented the suit from stretching when inflated. Gemini suits, however, were equipped with Dacron cord woven into a network of loose interconnecting links like fishnet throughout the suit. This lightweight yet extremely strong network was known as linknet.[2]

Underneath these suits, Gemini astronauts wore one-piece long underwear with five pockets to hold biomedical amplifiers. These were connected to body sensors (Figure 4.5) that relayed information to mission control about the astronaut's pulse, blood pressure, breathing, temperature, etc. The innermost layer of the space suit, the bladder layer, was made from rubberized nylon to hold air during pressurization. The David Clark Company's innovative Dacron linknet layer came next and effectively restrained the bladder layer without immobilizing the astronaut. The suit's outermost layer, made of aluminized coated nylon called Dupont HT-1 (High Temperature), functioned as an anti-snag garment, protecting the suit against the spacecraft instruments. The layers were bonded together only at the boots, zipper, biomedical plug, and the fittings for environmental controls.

In 1963, B. F. Goodrich accepted a NASA contract to test four David Clark pressure suits intended for use in the Gemini program. These were designated as G2C training and qualification pressure suits. Reliability of pressure-sealing zippers, visor closures, glove and helmet seals, and inlet/outlet fittings on the environmental control system were all tested at maximum safety pressures.

The basic concept of a single-walled pressure suit was retained for extravehicular activity. The David Clark Company experimented with a special cover layer which astronauts would don over the Gemini suit (Figure 4.6) before the astronauts lifted off, because the small Gemini spacecraft had no room for the men to modify their suits. AiResearch developed the first portable Gemini extravehicular life support system (ELSS) chest pack, which utilized a 7.5-m (25-ft.) "umbilical" assembly serving three purposes: Electrical wires provided communication and bioinstrumentation transmittal; a 453-kg (1,000-lb.) test tether attached the astronaut to the spacecraft, but allowed him to move around the spacecraft exterior; and a hose supplied spacecraft oxygen to the suit. A chest pack provided emergency oxygen from a high-pressure oxygen bottle. If the umbilical failed, the astronaut manually activated the emergency supply.

A second ELSS semi-open system was used on later Gemini missions. This ELSS provided increased thermal capacity from 140 Btu./hr. to peaks of 2,000 Btu./hr. High-pressure spacecraft oxygen flowed through an umbilical connected to the side of a chest pack in case of emergency. A selector valve permitted the astronaut to adjust flow rate. Exhaust gas was cooled by a water evaporation heat exchanger. Waters of respiration and perspiration were condensed in the heat exchanger and augmented the exchanger's water supply. Carbon dioxide was vented overboard. Emergency oxygen was automatically activated if the normal supply failed. The ELSS weighed 21.3 kg (47 lb.) and was stored in the spacecraft prior to launch.[3]

But there were problems. Despite these demanding specifications, the G2C pres-

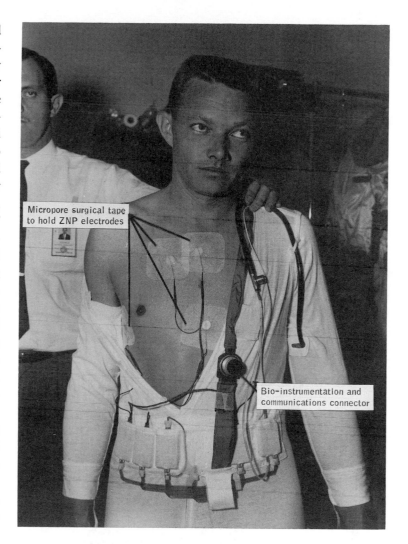

Micropore surgical tape to hold ZNP electrodes

Bio-instrumentation and communications connector

4.5

Gemini long underwear shows the easily attached bioinstrumentation and communication connections. (1965, NASA photo S-65-2484)

sure suit stretched out of shape during evaluation testing in the Gemini spacecraft engineering mock-up. A fixed helmet visor guard increased the height of the helmet to the point of interference with the spacecraft. David Clark took back the helmet and suit to be reworked.[4]

The improved David Clark prototype suit and helmet, redesignated G3C (Figure 4.7), were generally acceptable to the crew and compatible with the spacecraft. The suit's cover layer was made from uncoated HT-1 nylon, or Nomex material

4.6

Qualification and reliability testing of the G2C training suit began in early 1964. Here a NASA/MSC engineer demonstrates the prototype protective thermal cover layer worn over the conventional G2C for EVA. (SI photo 15844)

designed to protect the linknet restraint layer and gas container. Heat-resistant Nomex material was chosen for the G3C suit because it stretched less and did not flake or peel as did aluminized nylon.[5]

The Gemini G3C intravehicular space suit passed numerous development tests, including tests for suit performance, mobility, visibility, and long-term comfort. The communication system was integrated into the suit assembly for the first time. Once the astronaut connected his suit to the cabin's environmental control system, he had a pure oxygen supply at 34 kPa (5 psi), pressurization and humidity control, and the ability to remove strong odors and control his oxygen flow. The ventilation system, called an open-loop system, brought cool oxygen into the suit at chest level, carried it through the helmet, clearing the visor of any fog, and to the torso and upper and lower extremities, and then forced it out an exhaust line for purification and reuse. An inflight port inserted through the helmet wall allowed the astronaut to drink while in the pressurized suit. The helmet and gloves could be removed during normal spaceflight for eating and personal hygiene. Astronauts tested many urine collection devices (Figure 4.8) before a simple device was selected (Figure 4.9). Worn on the thigh, it was connected to the astronaut by rubber tubing.[6]

Since there was no planned EVA (walk-in-space), Gus Grissom and John Young each wore a simple G3C pressure suit aboard the first manned Gemini flight, *GT-3* (T for Titan rocket), in March 1965. The nearly five-hour flight proved the suit capable of sustaining life. There were no complications. Pulse, respiration, and blood pressure for both astronauts were in the expected ranges. The flight

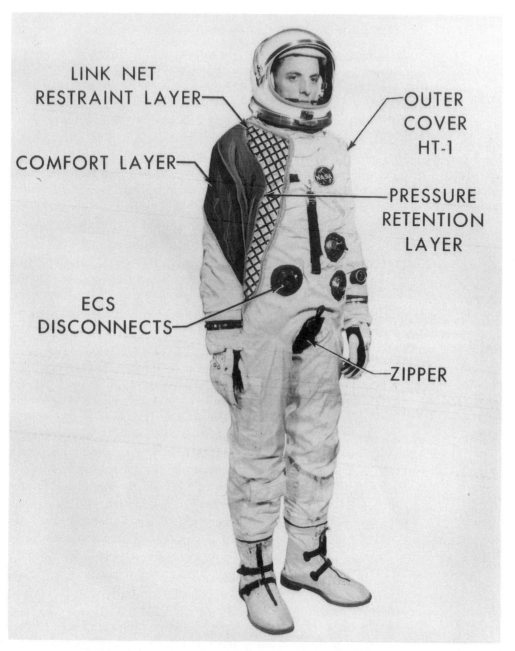

LINK NET
RESTRAINT LAYER

COMFORT LAYER

ECS
DISCONNECTS

OUTER
COVER
HT-1

PRESSURE
RETENTION
LAYER

ZIPPER

4.7
The first G3C flight suits were delivered to MSC in August 1964. This photo shows a cutaway of the suit, exposing the Dacron linknet restraint layer, a breakthrough in space suit development that greatly improved overall Gemini suit mobility. (NASA photo S-66-1847)

4.8 *(Top)*
Astronauts experimented with a number of different prototype urine transport systems, including these somewhat tortuous-looking creations. (1965, NASA photo S-65-3529)

4.9
This was the urine collection device selected: a simple contraption worn on the thigh and connected to the astronaut by tubing. (February 18, 1966, NASA photo S-66-1759)

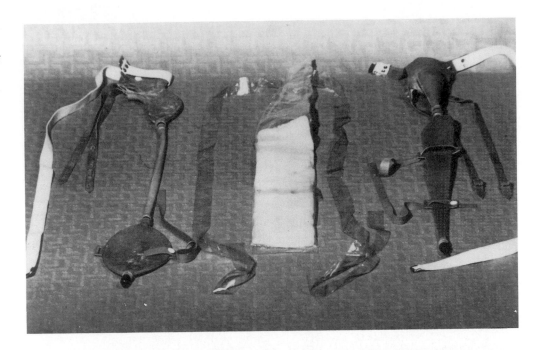

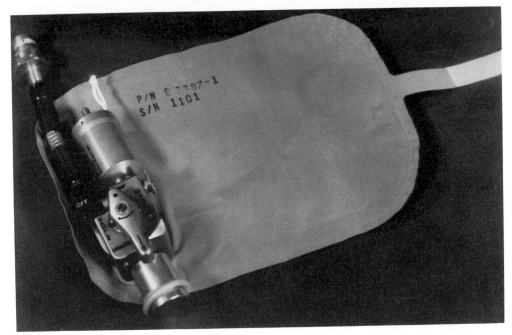

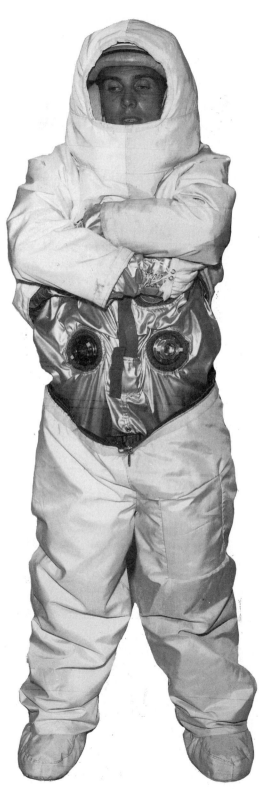

also obtained some scientific results. Grissom and Young found that they could effect minute changes in the orbital path of their spacecraft and manually control the reentry phase.

Extravehicular activities were planned for the second manned flight. Technicians tested an S1C (Figure 4.10), a prototype hazardous-environment suit, in simulated space conditions of −101° to 121°C (−150° to 250°F) for periods of approximately four hours and found it to be worthy. The S1C was actually modified from the G3C to provide the capability of working comfortably in the harsh space environment. From this point through the final mission of the Gemini program, the Gemini suit constantly changed as astronauts requested additional alterations.

GEMINI 4, 5, 6-A, AND 7

The outer layer of the next David Clark Gemini suit type, the G4C (Figure 4.11), was a follow-on version of the S1C hazardous-environment suit. The inner layers retained the design of the basic G3C suit. The cover layer, including heavier thermal and micrometeoroid protection, meant that the G4C EVA suit weighed 15.4 kg (34 lb.), which was 4.5 kg (10 lb.) more than the standard G3C suit.

Gemini astronauts wore cotton garments underneath their G4C suits. The torso assembly consisted of an oxford nylon liner (the innermost layer) fastened with Velcro tape to the gas container layer. The liner assembly helped diffuse the ventilation gas over the entire body through triloc material assemblies (similar to that used in the Mercury suit) located down the front of each leg, at the vent

4.10
NASA/MSC engineer wears the prototype S1C thermal- and micrometeoroid-protective overgarment on top of a conventional Gemini space suit. (1964, SI photo 83-15517)

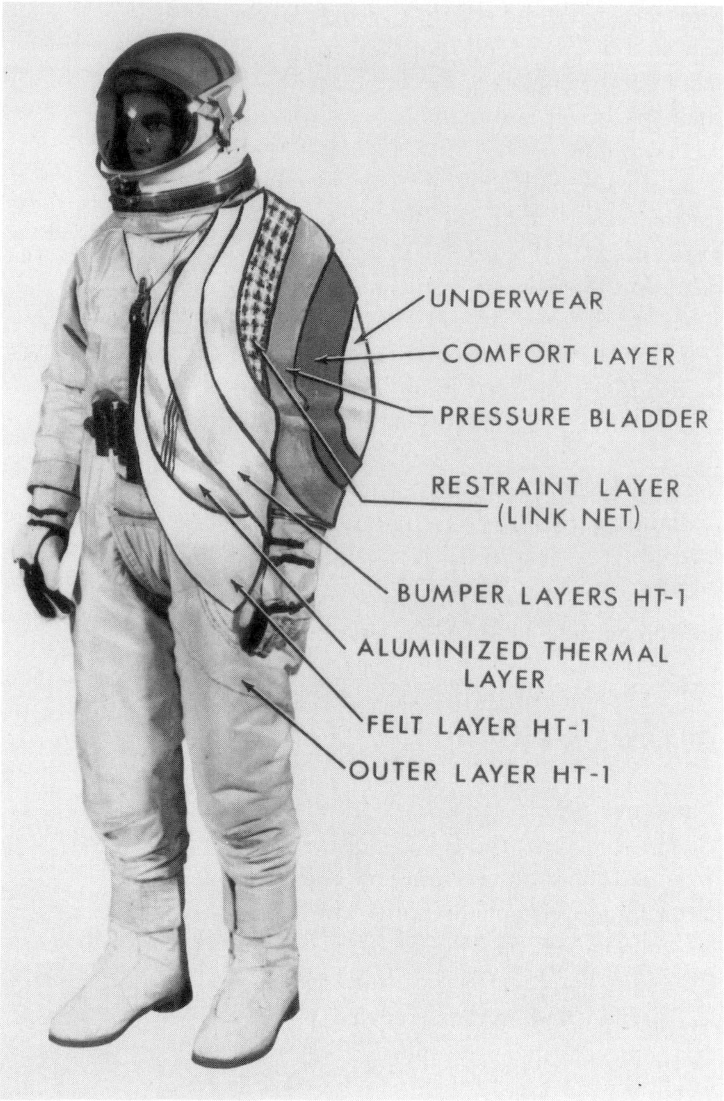

UNDERWEAR

COMFORT LAYER

PRESSURE BLADDER

RESTRAINT LAYER
(LINK NET)

BUMPER LAYERS HT-1

ALUMINIZED THERMAL
LAYER

FELT LAYER HT-1

OUTER LAYER HT-1

4.11

This cutaway illustration shows the basic five-layer G4C EVA space suit worn over underwear, with a two-layer micrometeoroid-protective cover layer. (May 1965, NASA photo S-65-4970)

exhaust, and leading from the exhaust area around the front and up the back to the neck. The gas container layer, fabricated from neoprene-coated ripstop nylon, was fastened with Velcro tape and snaps to the interior of the assembly. A Nomex linknet layer restrained the gas container to the contours of the body and allowed maximum mobility during suit pressurization.[7]

The cover layer of the G4C protected the restraint and gas container assembly and provided protection from thermal and micrometeoroid conditions during EVA. Micrometeoroids are small celestial bodies which move with considerable velocity because of the lack of resistance in space. They travel at speeds of up to 30 km/sec. (18 mi./sec.) and are so numerous that they could pose a real threat of depressurization. The temperature range which might be encountered during EVA depends largely on whether the astronaut leaves the spacecraft while it is on the sunward side or the dark side of its orbit. EVA suits were made to withstand temperatures between −60° and 164°C (−76° and 327°F).

This cover layer formed an integral part of the torso assembly, fabricated from a single outer and double inner cover of uncoated nylon over alternating Mylar and Dacron batten superinsulation, with a final layer of nylon projectile-resistant material. This was tested by firing .38-mm (.015-in.) Pyrex spheres into a torso suit and checking for penetration, perforation, and amount of leakage. These simulated micrometeoroids, traveling at an average velocity of 3,700 m/sec. (4,000 yd./sec.), caused only minute perforations and negligible leakage.

The helmet (Figure 4.12), fabricated from fiberglass and resin into a spherical dome with a Plexiglas visor, contained an anti-buffet liner and a ventilation-distribution system. An over-visor incorporated into the helmet design protected against glare, ultraviolet light, and extreme thermal conditions. A third over-visor provided impact protection.

G4C gloves (Figure 4.13) were designed for finger and wrist mobility, finger sensitivity, durability and abrasion protec-

tion, self-adjustment, and integrated finger and palm ventilation. The glove assembly had two individual layers: an inner glove, which was a molded gastight rubber bladder, and an outer glove fabricated of white nylon, which protected the inner glove. The gloves had fingertip lighting and an articulated palm restraint to prevent ballooning during pressurization. The boot assembly, designed as an attachment to the torso assembly, contained four layers: nylon, cotton-covered sponge, high-temperature fabric, and a multilayer protective cover.

The NASA/MSC Crew Systems Division experimented with reducing the bulk of the G4C space suit, because crews complained about discomfort when they were required to wear the EVA suits intravehicularly for prolonged periods. Tom Stafford wore a simplified G4C space suit aboard *GT-6A*, with only a single-layer cover layer, while Wally Schirra wore the

intravehicular G3C model. Both astronauts still experienced back discomfort, possibly from pressure points created by the long back zipper.

A basic suit worn over one-piece underwear consisted of five layers: lightweight blue oxford nylon, heavier blue oxford nylon, strong neoprene-coated nylon, linknet restraint material, and a cover layer of Nomex. Depending upon mission requirements, a choice of three different cover layers was possible. The command pilot's suit might be provided with one layer of mediumweight HT-1 uncoated Nomex. Ed White's EVA suit on *Gemini 4* contained two HT-1 nylon micrometeoroid-protection layers, seven aluminized Mylar layers alternating with seven layers of unwoven Dacron superinsulation, one HT-1 Nomex felt layer, and a final HT-1 Nomex outer protective layer. Ed White also wore another multilayer torso unit over the cover layer to give additional protection to the shoulders and arms.

The pilots of *Gemini 9* and *10* had suits that incorporated two layers of neoprene-coated nylon, seven layers of insulative fill (unwoven Dacron) alternating with seven layers of .00063-cm-thick (.00025-in.)

4.12 *(Top)*
The G3C helmet had a fiberglass shell with a Plexiglas visor and a liner inside for impact and shock absorption. It was worn with the G3C suit for intravehicular use. Note the inflight drinking port mounted through the helmet wall. (1965, NASA photo S-65-23033)

4.13
David Clark Company G3C gloves with wrist bearing disconnects were worn for intravehicular use. Tips of the second and third fingers contained tiny flashlight bulbs to help navigation during periods of darkness. (1965, NASA photo S-65-23034)

(6-oz.) nylon fabric. Overall size and bulk were considerably reduced from previous suits. A pressure-sealing closure at the neck permanently attached the fabric helmet to the torso assembly. When the helmet was in a partial-wear position, it was stowed in a fabric headrest case behind the head, readily available for quick donning. Bioinstrumentation, communications, and blood pressure entries were the same as those used for G3C and G4C suits.[11]

Lovell and Borman took off their G5C soft suits repeatedly during their 14-day flight. They reported that the removable suits were a great boon to comfort and recommended suitless flight for long-duration missions.

GEMINI 8, 9-A, 10, 11, AND 12

Modifications and improvements to space suits were made as necessary throughout the rest of the Gemini missions. The missions were not all uniformly successful. EVA was again included among the mission objectives for *Gemini 8* in March 1966. Neil Armstrong and Dave Scott experienced serious spacecraft problems, however, after their docking exercise with an Agena target vehicle, and they were forced to end their mission prematurely. Scott was equipped with a G4C EVA suit, while Armstrong wore the G3C.[12]

Gemini 9 was scheduled for launch of the augmented target docking adapter (ATDA) June 1, 1966. Computer problems delayed launch until June 3. Tom Stafford and Gene Cernan, now in *Gemini 9-A*, would attempt an ambitious EVA. Because Cernan would test an astronaut maneuvering unit (AMU) outside

the spacecraft, his suit was modified. The cover layer (Figure 4.16), identical to the standard EVA cover layer in the torso and arms, incorporated additional layers in the leg area to protect the wearer from the AMU's hydrogen peroxide thrusters when they fired. From the inside out, the suit's legs were made of two layers of neoprene-coated nylon, one layer of HT-1 uncoated nylon, 11 layers of insulative fill of fiberglass cloth alternated with aluminized H-film,* topped by a final layer of Chromel-R cloth (stainless steel). This protected the astronaut from surface temperatures of 650°C (1,200°F), while maintaining an internal temperature not higher than 43°C (110°F). The EVA portion of the cover layer weighed 1.3 kg/m² (56 oz./yd.²) as compared to .64 kg/m² (27 oz./yd.²) for the standard G4C cover layer material.[13]

Cernan's EVA activities were hampered because his body tended to rotate away from the spacecraft or work area, making each task more difficult to accomplish. Velcro pads slipped over his gloves (Figure 4.17) were of doubtful value in helping him maintain a proper body attitude. The exertion involved in just maintaining his body position produced a good deal of heat in his space suit, and at one point, he felt the back of his suit getting hot. The extra rotations experienced by Cernan as he tried to accomplish the experiments on his checklist increased his work load. As a result, he experienced excess moisture build-up inside his suit, visor fog, and fatigue. Stafford reported

---

*H stands for high temperature. H-film is a thin transparent film similar to Mylar. HT-1 is a high-temperature-resistant nylon material.

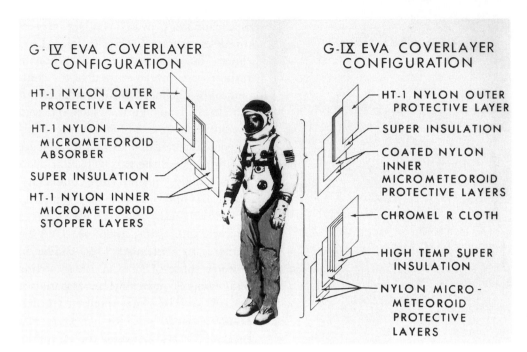

G·Ⅳ EVA COVERLAYER
CONFIGURATION

HT-1 NYLON OUTER
PROTECTIVE LAYER

HT-1 NYLON
MICROMETEOROID
ABSORBER

SUPER INSULATION

HT-1 NYLON INNER
MICROMETEOROID
STOPPER LAYERS

G·Ⅸ EVA COVERLAYER
CONFIGURATION

HT-1 NYLON OUTER
PROTECTIVE LAYER

SUPER INSULATION

COATED NYLON
INNER
MICROMETEOROID
PROTECTIVE LAYERS

CHROMEL R CLOTH

HIGH TEMP SUPER
INSULATION

NYLON MICRO-
METEOROID
PROTECTIVE
LAYERS

4.16
Gene Cernan wore this specially adapted G4C EVA suit June 3, 1966, during his GT-9A mission. It incorporated two layers of neoprene-coated nylon instead of the HT-1 nylon used on other Gemini suits. Suit designers also used a woven metal cloth called Chromel-R as an exterior covering on the leg areas to protect Cernan from the hot gases the astronaut maneuvering unit (AMU) was expected to generate. (SI photo 83-15507)

that communications through the AMU's trans-receiver were also garbled. The crew sadly decided against a flight test of the maneuvering unit and cut short Cernan's EVA. Postflight analysis of Cernan's suit supported their decision. The thermal layer began to rip in the area of the super-insulation, allowing solar radiation to penetrate which contributed to the over-heating problem.[14]

John Young and Mike Collins were assigned EVA during their *Gemini 10* mission in July 1966. Collins' G4C cover layer consisted of two layers of neoprene-coated nylon, seven of insulative fill of unwoven Dacron alternating with seven of aluminized Mylar, and a final layer of HT-1 uncoated nylon, weighing in at .64 kg/m² (27 oz./yd.²). The G4C cover layers for the command pilots of *Gemini 9, 10, 11,* and *12,* who did not actually participate in EVA, contained one light-weight layer of .14 kg/m² (6 oz./yd.²) HT-1 uncoated nylon.[15]

Gemini 10 astronauts opened the hatch for EVA on three different occasions. Stand-up EVA (the astronaut does not actually leave the spacecraft but stands in the open hatch area) began 23 hours into the flight. A nylon tether served as a safety restraint during this activity, connecting the wearer to the spacecraft. Collins noted in *Carrying the Fire* that stand-up EVA was much like sticking your head up through the roof of a fast-moving car—exciting and not at all unpleasant. In this position, the crew performed experiments involving photography, star navigation, and space-radiation measurements. Collins began tethered EVA with a 15-m (50-ft.) umbilical and a hand-held maneuvering unit. The maneuvering unit worked well in transferring Collins between the spacecraft and the Agena target vehicle,

4.17

Velcro pads on the gloves and boots of the EVA suit worn on *Gemini 9* in June 1966 were intended to help the astronaut maintain an upright position, but were of doubtful value. (SI photo 83-15839)

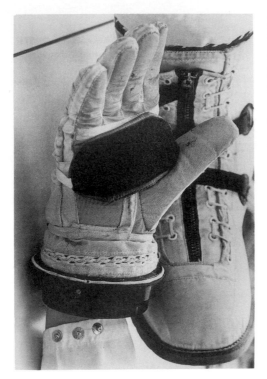

but the umbilical proved difficult to manage. NASA suggested using a shorter one of 9 m (30 ft.) for future flights.[16]

Gemini 10 suits were modified as follows: The arms and legs were removed from the underwear at the torso seam, which helped keep the astronauts cool; the EVA sun visor used Velcro straps to interface with mating Velcro attached to the helmet rather than the aluminum locking mechanism; fingertip lights contained red lenses to prevent overexposing film during dark-side photographic experiments; and an anti-fog visor kit containing a wetting agent was added for inflight and EVA use.[17]

Helmets for the astronauts who crewed *Gemini 9* through *12* were constructed from multiple layers of epoxy-impregnated fiberglass cloth with a polycarbonate pressure-sealing visor. The polycarbonate provided impact protection far exceeding that of Plexiglas. EVA helmets incorporated a polycarbonate pressure visor similar to that of the command pilot's helmet, with the addition of a single-layer sun visor, which replaced the multi-lens visor previously used. Impact protection came from the pressure visor rather than the extravehicular over-visor assembly, thus reducing complexity, weight, and bulk.[18]

Gemini 11 astronauts Charles Conrad and Richard Gordon completed EVA assignments in September 1966 similar in nature to those of mission 10. In setting up a camera, retrieving an experiment package, and attaching a tether, Gordon became fatigued and further EVA was canceled. He also collected an accumulation of perspiration in one eye from his exertions. Conrad and Gordon believed that they had practiced too much for their EVA. They were ready two hours ahead of schedule and felt this contributed to Gordon's becoming overheated and tired. He also had difficulty putting on the gold-plated EVA visor. Stand-up EVA performed later in the mission went as planned, and no significant problems were experienced with the suits.[19]

During the last Gemini flight in November 1966, *Gemini 12* command pilot Jim Lovell wore the standard G4C suit, but Buzz Aldrin donned a G4C fitted with a modified extravehicular cover layer. The termal cover layer was quilted to the first layer of micrometeoroid-protective material in a rectangular pattern over the torso to prevent the aluminized H-film or Mylar from tearing. The crew made use of their anti-fog kits prior to EVA. The inside of the visor was treated with an anti-fog compound. Additionally, the helmet had a reservoir, or plenum area in its shell, perforated to effect a defogging airflow when connected to the main suit assembly and the oxygen system is actuated. Aldrin was able to attach a tether to the spacecraft docking bar, evaluate a restraint system, and dock and undock from the Gemini Agena target vehicle—all without great fatigue or problems with his Gemini suit.[20]

Gemini extravehicular activity missions proved that humans could perform useful tasks in space, install and retrieve scientific equipment, and move from vehicle to vehicle. All crew members had problems maintaining body stability while working in a space suit, and this prevented the achievement of some complex EVA tasks. Each mission built upon lessons learned during the previous one (Fig-

ures 4.18, 4.19, and 4.20), and Aldrin's EVA during the last flight proved more successful because he utilized new portable handrails, foot restraints, and waist tethers.

For Apollo missions, engineers would work to provide a more stable work platform for EVA participants, as bioengineers sought to improve spacecraft and portable life support systems. Life scientists would tackle the problem of excessive metabolic expenditures during EVA. And manufacturers, NASA engineers, and astronauts would work together to simplify space suit donning and doffing and to improve suit mobility, a critical characteristic for crews assigned to explore the surace of the moon.

4.18 (Below), **4.19** (Page 70), **and 4.20** (Page 71) Diagrams illustrate type and function of Gemini suits used for each flight during March 1965 through November 1966. (illustrations courtesy of Dennis Gilliam, SI photos 92-648, 92-649, and 92-650)

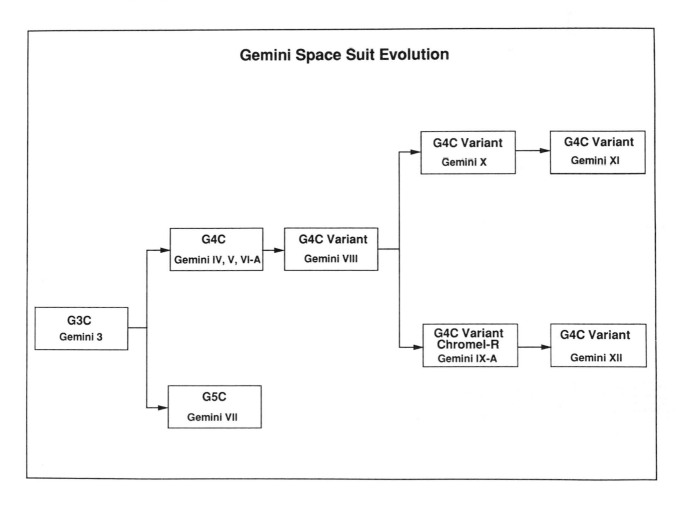

Gemini Space Suit Evolution

Gemini Suit Evolution

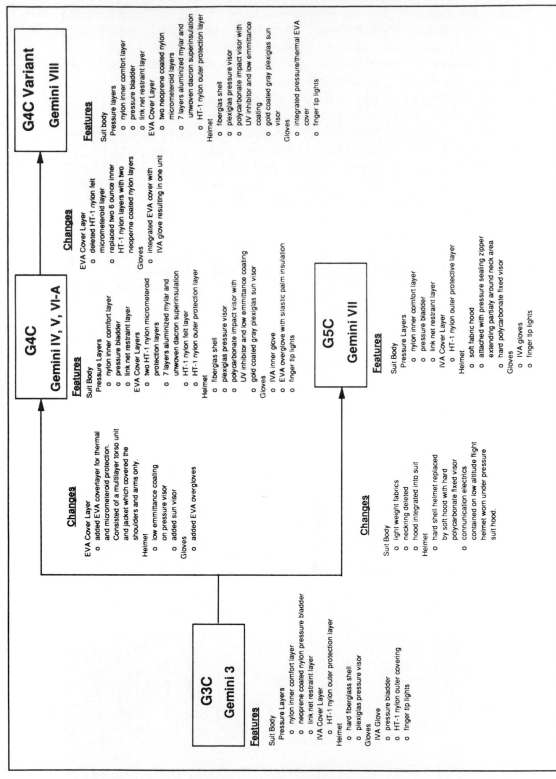

G3C — Gemini 3

Features

Suit Body
Pressure Layers
- o nylon inner comfort layer
- o neoprene coated nylon pressure bladder
- o link net restraint layer

IVA Cover Layer
- o HT-1 nylon outer protection layer

Helmet
- o hard fiberglass shell
- o plexiglas pressure visor

Gloves
IVA Glove
- o pressure bladder
- o HT-1 nylon outer covering
- o finger tip lights

Changes

EVA Cover Layer
- o added EVA coverlayer for thermal and micrometeroid protection. Consisted of a multilayer torso unit and jacket which covered the shoulders and arms only.

Helmet
- o low emmittance coating on pressure visor
- o added sun visor

Gloves
- o added EVA overgloves

G4C — Gemini IV, V, VI-A

Features

Suit Body
Pressure Layers
- o nylon inner comfort layer
- o pressure bladder
- o link net restraint layer

EVA Cover Layers
- o two HT-1 nylon micrometeroid protection layers
- o 7 layers aluminized mylar and unwoven dacron superinsulation
- o HT-1 nylon felt layer
- o HT-1 nylon outer protection layer

Helmet
- o fiberglas shell
- o polycarbonate impact visor with UV inhibitor and low emmittance coating
- o gold coated gray plexiglas sun visor

Gloves
- o IVA inner glove
- o EVA overglove with silastic palm insulation
- o finger tip lights

Changes

EVA Cover Layer
- o deleted HT-1 nylon felt micrometeroid layer
- o replaced two 6 ounce inner HT-1 nylon layers with two neoperne coated nylon layers

Gloves
- o integrated EVA cover with IVA glove resulting in one unit

G4C Variant — Gemini VIII

Features

Suit body
Pressure layers
- o nylon inner comfort layer
- o pressure bladder
- o link net restraint layer

EVA Cover Layer
- o two neoprene coated nylon micrometeroid layers
- o 7 layers aluminized mylar and unwoven dacron superinsulation
- o HT-1 nylon outer protection layer

Helmet
- o fiberglas shell
- o plexiglas pressure visor
- o polycarbonate impact visor with UV inhibitor and low emmittance coating
- o gold coated gray plexiglas sun visor

Gloves
- o integrated pressure/thermal EVA cover
- o finger tip lights

Changes

Suit Body
- o light weight fabrics
- o neckring deleted
- o hood integrated into suit

Helmet
- o hard shell helmet replaced by soft hood with hard polycarbonate fixed visor
- o connunication electrics contained on low altitude flight helmet worn under pressure suit hood.

G5C — Gemini VII

Features

Suit Body
Pressure Layers
- o nylon inner comfort layer
- o pressure bladder
- o link net restraint layer

IVA Cover Layer
- o HT-1 nylon outer protective layer

Helmet
- o soft fabric hood
- o attached with pressure sealing zipper extending partialy around neck area
- o hard polycarbonate fixed visor

Gloves
- o IVA gloves
- o finger tip lights

Gemini Suit Evolution

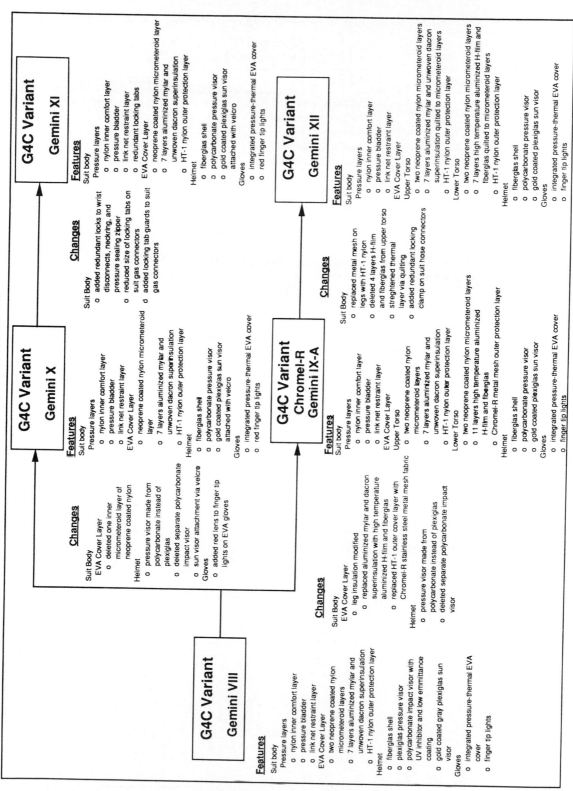

G4C Variant — Gemini VIII

Features

Suit body
- Pressure layers
  - nylon inner comfort layer
  - pressure bladder
  - link net restraint layer
- EVA Cover Layer
  - two neoprene coated nylon micrometeroid layers
  - 7 layers aluminized mylar and unwoven dacron superinsulation
  - HT-1 nylon outer protection layer
- Helmet
  - fiberglas shell
  - plexiglas pressure visor
  - polycarbonate impact visor with UV inhibitor and low emmittance coating
  - gold coated gray plexiglas sun visor
- Gloves
  - integrated pressure-thermal EVA cover
  - finger tip lights

Changes (Gemini VIII → Gemini X)

Suit Body
- EVA Cover Layer
  - deleted one inner micrometeroid layer of neoprene coated nylon
- Helmet
  - pressure visor made from polycarbonate instead of plexiglas
  - deleted separate polycarbonate impact visor
- Gloves
  - added red lens to finger tip lights on EVA gloves

G4C Variant — Gemini X

Features

Suit body
- Pressure layers
  - nylon inner comfort layer
  - pressure bladder
  - link net restraint layer
- EVA Cover Layer
  - neoprene coated nylon micrometeroid layer
  - 7 layers aluminized mylar and unwoven dacron superinsulation
  - HT-1 nylon outer protection layer
- Helmet
  - fiberglas shell
  - polycarbonate pressure visor
  - gold coated plexiglas sun visor attached with velcro
- Gloves
  - integrated pressure-thermal EVA cover
  - red finger tip lights

Changes (Gemini X → Gemini XI)

Suit Body
- added redundant locks to wrist disconnects, neckring, and pressure sealing zipper
- reduced size of locking tabs on suit gas connectors
- added locking tab guards to suit gas connectors

G4C Variant — Gemini XI

Features

Suit body
- Pressure layers
  - nylon inner comfort layer
  - pressure bladder
  - link net restraint layer
  - redundant locking tabs
- EVA Cover Layer
  - neoprene coated nylon micrometeroid layer
  - 7 layers aluminized mylar and unwoven dacron superinsulation
  - HT-1 nylon outer protection layer
- Helmet
  - fiberglas shell
  - polycarbonate pressure visor
  - gold coated plexiglas sun visor attached with velcro
- Gloves
  - integrated pressure-thermal EVA cover
  - red finger tip lights

G4C Variant — Chromel-R — Gemini IX-A

Features

Suit body
- Pressure layers
  - nylon inner comfort layer
  - pressure bladder
  - link net restraint layer
- EVA Cover Layer
  - Upper Torso
    - two neoprene coated nylon micrometeroid layers
    - 7 layers aluminized mylar and unwoven dacron superinsulation
    - HT-1 nylon outer protection layer
  - Lower Torso
    - two neoprene coated nylon micrometeroid layers
    - 11 layers high temperature aluminized H-film and fiberglas
    - Chromel-R metal mesh outer protection layer
- Helmet
  - fiberglas shell
  - polycarbonate pressure visor
  - gold coated plexiglas sun visor
- Gloves
  - integrated pressure-thermal EVA cover
  - finger tip lights

Changes (Gemini VIII → Gemini IX-A)

Suit Body
- EVA Cover Layer
  - leg insulation modified
    - replaced aluminized mylar and dacron superinsulation with high temperature aluminized H-film and fiberglas
    - replaced HT-1 outer cover layer with Chromel-R stainless steel metal mesh fabric
- Helmet
  - pressure visor made from polycarbonate instead of plexiglas
  - deleted separate polycarbonate impact visor

Changes (Gemini IX-A → Gemini XII)

Suit Body
- replaced metal mesh on legs with HT-1 nylon and fiberglas from upper torso
- deleted 4 layers H-film
- strenghtened thermal layer via quilting
- added redundant locking clamp on suit hose connectors

G4C Variant — Gemini XII

Features

Suit body
- Pressure layers
  - nylon inner comfort layer
  - pressure bladder
  - link net restraint layer
- EVA Cover Layer
  - Upper Torso
    - two neoprene coated nylon micrometeroid layers
    - 7 layers aluminized mylar and unwoven dacron superinsulation quilted to micrometeroid layers
    - HT-1 nylon outer protection layer
  - Lower Torso
    - two neoprene coated nylon micrometeroid layers
    - 7 layers high temperature aluminized H-film and fiberglas quilted to micrometeroid layers
    - HT-1 nylon outer protection layer
- Helmet
  - fiberglas shell
  - polycarbonate pressure visor
  - gold coated plexiglas sun visor
- Gloves
  - integrated pressure-thermal EVA cover
  - finger tip lights

Table 4.1
Gemini G4C Materials
Pressure Garment Assembly (PGA)

Layer*	Material	Function
1	HT-1 uncoated nylon	Abrasion/flame resistance
2	HT-1 uncoated Nomex	Fire protection
3,5,7,9,11,13,15	Unwoven Dacron	Insulative fill spacer
4,6,8,10,12,14,16	Aluminized Mylar	Reflective insulation
17,18	Neoprene-coated nylon	Inner liner
19	Linknet	Restraint
20	Strong neoprene-coated nylon	Bladder layer for pressurization
21	Heavier blue oxford nylon	Restraint
22	Lightweight blue oxford nylon	Comfort

*The 22 layers of protective materials listed here are representative of those used in Gemini space suits. The exact number of layers varied with the suit's function, i.e., extravehicular or intravehicular.

CHAPTER 5

Apollo
Space Suits

It is extraordinary that the unattainable dream of centuries—travel to the moon—became a reality only some 66 years after the first powered flights by the Wright brothers. The 20 years since those epic Apollo moon landings have even seen us become a little jaded about space travel. Yet during the late 1960s and the 1970s, the Apollo flights excited international public attention and enthusiasm unequaled by almost any other contemporary enterprise.

During the 24 months between the final Gemini mission and the first earth-orbital manned test of the Apollo command and service module (CSM) in October 1968, NASA worked to ensure the safety and reliability of its two quite different Apollo spacecraft. Each lunar landing mission involved an Apollo CSM which would rendezvous and dock with a lunar module (LM) shortly after launch, the two vehicles having been carried aloft in a stack by the same Saturn V launch vehicle. In the docked configuration, the three Apollo astronauts would journey toward and begin orbit around the moon. While in orbit, two astronauts would transfer to the lunar landing module and disengage the two craft, leaving one crewman behind to monitor the landing activities from the CSM.

To power the Apollo journey to the moon, Wernher von Braun suggested clustering a number of smaller rockets. The von Braun team's Juno V gained support and, after

● ●

Costs in Perspective

The 1960s were turbulent years. The war in Southeast Asia continued to escalate. By the end of 1970, this war cost the United States $108.5 billion. The achievements of Apollo were clouded over by the program's high costs, as well as those of the war in Indochina. Demonstrations against the war and also for social change became commonplace. Rioting erupted in many cities. It was difficult to justify allocating money for space exploration when basic needs of food, shelter, and clothing were not being met for so many Americans. Robert Kennedy and Martin Luther King were assassinated in 1968, and President Johnson's plans for correcting social ills fell far short of his goals.

**   The entire Apollo lunar program including research and development cost $25.5 billion. By comparison, in 1972, $71.7 billion went to the Department of Health, Education, and Welfare and $70.5 billion to the Department of Defense. That year NASA received $3.31 billion. In the 14 years between 1959 and 1972, NASA received a total of $46.8 billion. People must realize $46.8 billion did not go up into space; men and machinery did. Much of that money went right back into the economy. It provided jobs for steel and other allied blue collar workers, engineers, and the many specialized staff of the space program. Few people acknowledge how little the American space program has really cost.[1]**

● ●

renaming, became the first in the Saturn family of rockets. Powerful F-1, H-1, and J-2 engines were developed to power ever larger Saturns. Eight H-1 rocket engines powered the Saturn 1B. Saturn 1Bs were used to launch *Apollo 7* and *Apollo-Soyuz*. Five F-1 engines, producing as much power as five Saturn 1s lashed together, formed the first stage of the Saturn V. The S-II formed the second stage powered by five J-2 engines. The third stage was the S-IVB powered by a single J-2 engine. It was above this Saturn V that the comparatively tiny Apollo spacecraft was launched toward the moon.

Apollo crew members, like Mercury and Gemini astronauts before them, were provided with protective suits for the launch, docking, and reentry phases of their missions and for extravehicular activity, which would now include lunar ac-

tivity. Unlike their predecessors, Apollo astronauts, had reasonable living space in their command modules and could work in the relative comfort of lightweight intravehicular garments whenever bulkier space suits were not required.

But there were many extra requirements for the Apollo pressure suit besides protecting the wearer if the spacecraft cabin depressurized suddenly. Lunar temperatures between −150° and 120°C (−238° and 248°F), cosmic radiation,* and exposure to high-speed meteoroid particles in a gravity one-sixth that of earth's would expose the astronauts to the harshest environment yet encountered by

---

*Cosmic radiation is energy emitted in space in the form of waves or particles.

humans. The encapsulated environment provided by their space suits had to guarantee their safety.

During extravehicular activity, Apollo astronauts could not be hampered with life-supporting umbilicals connecting them to the spacecraft. Therefore, their pressure garment assemblies (space suits) had to be compatible with a portable oxygen supply/life support system, worn like a backpack and holding sufficient oxygen to support lengthy EVAs. Exploring the moon would require the astronaut to walk over rugged terrain in his pressurized suit, to pick himself up unaided should he stumble, and to perform work requiring manual dexterity, all of which called for a light, flexible suit sturdy enough not to tear.

A system for eating and drinking under zero gravity conditions while suited was called for, as well as a mechanism for communication. Provision also had to be made to collect and store or, if possible, eliminate body wastes. It was believed to be especially important that the suit be easy to don and doff without assistance in the confines of the Apollo spacecraft. In practice, however, the crew helped each other.[2]

DEVELOPMENT OF THE APOLLO SUIT

In 1961, United Technologies Hamilton Standard of Windsor Locks, Connecticut, was named by NASA as the prime contractor for the Apollo pressurized garment assembly. NASA assigned the company overall management responsibility of the suit program, as well as the development of the life support backpack required for lunar exploration. International Latex Corporation (ILC) was chosen to fabricate the suit, as a subcontractor, with Republic Aviation Corporation furnishing human factors evaluations. Republic was capable of testing ILC's suits in its own space simulation chamber, which could duplicate conditions of altitudes up to 480 km (300 mi.). In the early 1960s, Republic had created the experimental "tripod tepee" suit which offered an interesting basic solution to the problem of fatigue caused by heavy suits (Figure 5.1). The only drawback would have been getting easily in and out of that suit inside the spacecraft.

NASA selected Hamilton Standard and ILC from a field of 15 candidates (Figure 5.2) in a competition for a lunar life support/protection system. Both had entered space suits and portable life support systems (PLSS).[*] Some NASA evaluations favored ILC's suit but not its PLSS, while others favored Hamilton Standard's PLSS but not its suit. ILC's suit met all the agency's requirements and thus the company won the prime contract from NASA for its development in June 1962. Although Hamilton Standard lost the competition, it retained overall responsibility, as well as authority, for development of the PLSS. The PLSS was to contain a heat exchanger, an oxygen supply system, and a two-way radio for communications, with the option of a propulsion system that might be built into the pack.[3]

NASA directed that the development of the space unit should be conducted in three phases. A1C Block I "soft suits" (Figure 5.3) were similar to those worn by

---

[*]Actually, Hamilton Standard (PLSS) originally teamed with the David Clark Company (suit), and ILC (suit) teamed with Westinghouse (PLSS).

5.1
Republic Aviation pro-
duced an early concept
for a ''moon walk'' suit.
The astronaut would be
able to extend the legs of
the tripod tepee suit and
rest, seated on a little
built-in platform. (photo
courtesy of the American
Institute of Aeronautics
and Astronautics, SI
photo 88-32)

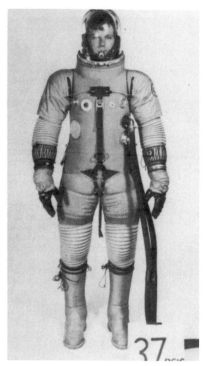

Gemini astronauts but modified to be compatible with interfacing hose connections in the Apollo spacecraft. These suits were to be used for earth-orbital test missions during which there would be no extravehicular activity. Block II soft suits, with thermal coveralls and special insulative material, were to be developed by ILC for lunar exploration tasks. The third phase called for an advanced design for missions of about 14 days on the lunar surface. The designers of these space suits followed the so-called "hard suit" approach, referring to the "lobster shell" restraint suit made by ILC for testing at NASA (Figure 5.4). The designs were for an exoskeletal (hard-shelled), anthropomorphic (human-shaped) suit that was articulated at the joints.[4]

ILC used a natural rubber in the joints of its first Block I prototype space suits,

which were dubbed state-of-the-art (SOA-IL) suits. Their prime objective at this point was to achieve maximum suit mobility. Designers were concerned, however, about the stability of rubber in a space environment.

Exposure to ultraviolet and infrared rays would eventually cause degradation of rubber, so ILC and its suppliers experimented with several candidate materials. They discovered that a film of Mylar or aluminum would provide a thermal barrier that offset radiated heat. But the SOA-IL suit was still too broad across the shoulders for the Apollo spacecraft, so the decision was made that the Block I suits would revert to a modified Gemini-type suit (A1C) produced by the David Clark Company.

Hamilton Standard reported that it intended to cancel ILC's contract in early

5.2

Three companies, from an original group of 15, emerged as serious contenders for the NASA contract for Apollo Block II (lunar mission) suits. The Hamilton Standard/ Goodrich design is shown on the left. The International Latex design, the winner, is in the middle, and the David Clark proposal is shown at right. (1965, SI photo 87-6757)

5.3

Left to right: Astronauts Eisele, Schirra, and Cunningham wear A1C Block I suits (similar to Gemini suits and also made by the David Clark Company) for mission simulator tests, June 1967. The A1C Gemini style helmet had been modified by the addition of a protective shell for the visor. (NASA photo 67-H-971)

1965 because of cost overruns, late deliveries, and poor performance of its prototype suits. The Manned Spacecraft Center concurred and Hamilton Standard began in-house work on its own suit with the aid of B. F. Goodrich. The David Clark Company also received a contract for the backup development of an Apollo Block II suit. ILC did not receive a contract to continue development of its own space suit, but the company was allowed to submit an entry for NASA's consideration, with development work to be paid for by ILC. The suits of all three companies were tested in June 1965. After an extensive test and evaluation program, NASA concluded that ILC's AX5L best met the requirements for the Block II lunar suit, having better mobility and a more compact pressurized shape.[5]

Hamilton Standard retained the contract to manufacture the portable life support system, but NASA's Manned Spacecraft Center in Houston assumed management of the total system. ILC agreed to develop the Apollo Block II suit under separate contract, and David Clark still had its contract to produce the Apollo Block I earth-orbital suits and a backup Block II design.

Apollo Block I suits, designated A1C (Figure 5.3), were similar to Gemini suits

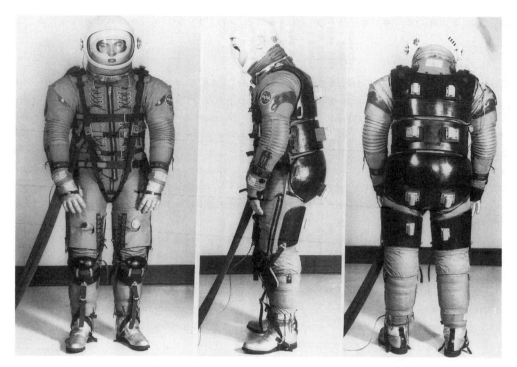

5.4
International Latex produced the "lobster shell" restraint suit for NASA's Ames Research Center. Designed for use during reentry, the restraint harness was removable and could be stowed during flight. (1966, NASA photo 66-H-180)

with minor modifications. The A1C helmet also had a protective shell for the visor. These suits were originally constructed of either an uncoated or an aluminized Nomex high-temperature-resistant material. However, after the tragic fire that claimed the lives of the first three Apollo astronauts at the launch site in January 1967, the suits were redesigned to include new fire-resistant materials. Rescue teams experienced great difficulty in extricating the bodies of the astronauts from their spacecraft because the suits had fused. Fed by pure oxygen in the spacecraft, white-hot flames, "like an acetylene torch," had raised the temperature beyond the melting point of the "high-temperature-resistant material."

Nomex, the nylon-type material used in this suit, has a melting point of 370°C (700°F).[6] The Apollo 204 Review Board,

assigned to investigate the accident, recommended that non-flammable materials replace combustible ones wherever possible. Nonmetallic materials were carefully screened. Fabrics, fasteners, film, and foams required thorough investigation. Nylon cloth in the spacecraft and in the suits was replaced by Beta cloth, a substance developed by Frederick S. Dawn's research team in conjunction with the Dow-Corning Company. Technically called Beta-silica fiber, it was a different material than that used in trade-name Fiberglas products. Beta-silica fibers could be spun into thin threads and then woven into fabric with a melting point of over 650°C (1,200°F) that would neither ignite nor produce toxic fumes.[7]

The Block II suits for Apollo lunar missions had to be reinforced against micrometeoroids, but still retain enough mo-

●　●　●　●　●　●　●　●　●　●　●　●　●　●　●　●　●　●　●

The ILC Story

Bruce Ferguson, former Public Relations Director of ILC Industries, Inc., Dover, Delaware, reminisced about his career and the Apollo space suits. When he joined the International Latex Corporation, Government and Industrial Division in 1966, there had been approximately 350 employees. In two years, the company grew to 1,400. ILC was looking for people with open minds. The concept of a walking suit that would provide mobility to get up after falling, perform useful tasks, and so on, was brand new. Space suit production needed people to sew and to cement (glue) pieces of material together to tolerances not achieved before. It needed people who could design and use patterns in entirely new and ingenious ways.

There were many sewing operations in the Dover area, dress and luggage manufacturers. Women from these industries were hired by ILC (Figures 5.5 and 5.6). They learned to read blueprints, to make evaluations, and to suggest improvements to the final design. The Apollo suit was, in reality, 21 suits superimposed one over another. Some layers were so thin that a piece of the material dropped from shoulder height might take five to ten seconds to hit the floor. This nonwoven Dacron provided a trilubricant for the aluminized Mylar layers, giving it a slippery mobility.

Special compounds were needed to produce liquid latex, the material used for the convolutes in the suit. This seemed like one of the strangest, mystical types of alchemy (reminding engineers of medieval days when sorcerers tried to turn base metals into gold). ILC hired senior technicians with many years of experience in the compounding and dipping processes from International Latex (the parent company). These men and women were skilled in the art of mixing and using the compounds. ILC engineers had tried to do the same thing, but their products didn't work. The accordion-shaped layer was built up by being repeatedly dipped in the latex compound in a process similar to candle-making. These convolutes were then reinforced with nylon fabric, tapes, and steel cables (Figure 5.7); they would be used on all Apollo, Skylab, and ASTP space suits. The convolute layer retained the pressurized air but still allowed mobility (Russell Colley's tomato-worm effect) and constant volume.

Initially, ILC had been a subcontractor to Hamilton Standard. ILC was to provide the articulation—the arms and legs—for the suit Hamilton Standard was building. A falling out between the companies caused ILC to be removed from the contract. Dr. Nisson A. Finkelstein (then president of ILC's Government and Industrial Division) threatened to sue both Hamilton Standard and NASA because he felt that ILC had lived up to its end of the contract.

●　●　●　●　●　●　●　●　●　●　●　●　●　●　●　●　●　●　●

• •

Instead of suing, ILC accepted NASA's offer to enter the contest independently with a company-funded design. Dr. Finkelstein went to the parent company and received about $110,000 to construct a suit. They had five weeks. Ferguson said the people at ILC Industries (and he meant the women) worked 24 hours a day for those five weeks, making the AX5L prototype.

Despite this, they entered the three-week competition a week late. A technician In the Hamilton Standard suit (A5H) was practicing climbing a ladder when the pressure blew his helmet off almost taking his nose with it. Mobility in the David Clark modified Gemini suit was so restrictive, it was impossible to use. ILC won the competition.

The three competition suits had been laid out for Vice President Hubert Humphrey's visit to the center. Two suits displayed NASA patches; ILC had not been given any so they devised their own. He asked why the ILC suit did not have a NASA patch. There was a long silence. "Well," he said, "I guess Brand X suit won the contest, didn't it?"

Dr. Finkelstein soon received a phone call from NASA/Houston inviting him to discuss a contract. Finkelstein said ILC was busy having a victory party. The highlight would be when he took the suit and threw it on the fire! The NASA man said, "You are kidding, aren't you?" "Partially," Finkelstein answered, "you really treated us badly, you know."

Because they had entered the competition independently, ILC owned the winning suit. They were now able to renegotiate the whole contract. ILC ended up with a cost-plus-fee contract, which is the normal way to do things in a new discipline. NASA bought the suit and it became the prototype model. Seamstresses took the suit apart and remade it many times until they developed the next stage, the A6L. Its design incorporated changes recommended by the Apollo 204 Review Board, including the addition of an outer layer of Beta fabric with metallic Chromel-R cloth patches.

When Roger Chaffee, Ed White, and Gus Grissom died in the 1967 Apollo fire, the whole program ground to a stop. (They had worn AICs, modified Gemini suits.) While NASA worked out its problems, ILC continued to develop their suit. The A7L was the follow-on suit: the first integrated thermal micrometeoroid model. Ferguson remembered how the suit had been tested on the Dover High School football field. "We pressurized the suit, tethered the technician to a long umbilical oxygen line. He fell down, got up, passed, ran, caught balls. We took the suit to NASA. We knew they could not refuse this suit. They bought it."

• •

5.5 (*Top*)
Michelle Tice works on a
sizing insert. (1968, photo
courtesy of ILC Dover,
Inc.)

5.6
Julia Brown works on a
glove restraint. (1968,
photo courtesy of ILC
Dover, Inc.)

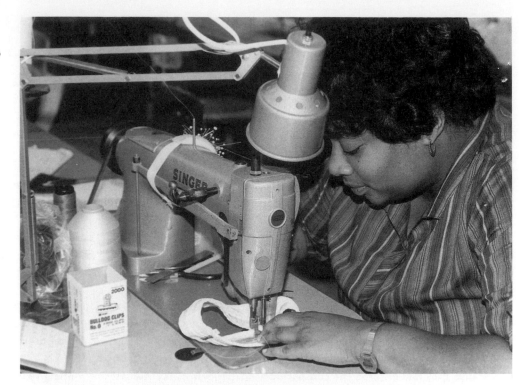

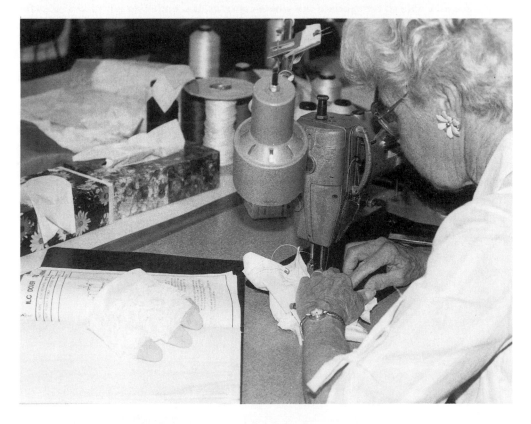

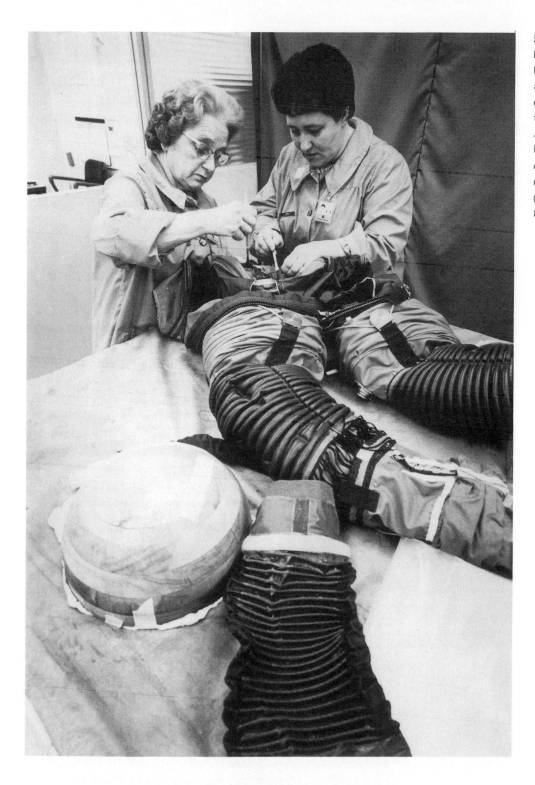

5.7
Delema Austin (*left*) and Delores Zeroles (who also worked on Shuttle-era suits at ILC) work on the fabrication of the Apollo constant volume bellows-like joints, which could follow astronauts' natural movements. (1968, SI photo 83-15837)

5.8
The liquid cooling garment allows water to circulate through an interlacing of small tubes stitched onto nylon spandex. This tubing, which rests nearly on the astronaut's skin, cools by direct conduction, almost eliminating perspiration. (1968, SI photo 83-15819)

bility and comfort to allow tasks to be accomplished on the lunar surface. Astronauts needed dexterity and tactility in gloved hands to perform experiments, assemble and use complex equipment and tools, and gather lunar rocks, all in a space vacuum. They had to be able to walk many miles on the lunar surface with a minimum of effort. The complete Apollo lunar suit system was, in fact, officially designated the extravehicular mobility unit (EMU) by NASA.

EXTRAVEHICULAR MOBILITY UNIT

The initial Apollo EMU configuration (manufactured by ILC) comprised several components: a constant wear garment (CWG), a liquid cooling garment (LCG) (Figure 5.8), a pressure garment assembly (PGA), an external thermal garment and meteoroid protection garment—later incorporated into an integrated thermal micrometeoroid garment (ITMG), the portable life support system (PLSS), the emergency oxygen system (EOS), and the waste management system (Figure 5.9).

Space suit designers had had many problems controlling the Mercury and Gemini air-cooled suits. The work loads of Apollo astronauts operating on the lunar surface would be much greater than any experienced during a typical Gemini extravehicular activity, and a more reliable cooling system was required. Devised with the assistance of D. R. Burton of the Royal Aircraft Establishment, Farnborough, England, the liquid cooling garment became the innermost layer of the EMU. Development of this garment followed a Farnborough study of air-cooled suits, experiments with a British prototype suit provided to NASA, and

● ●

Safety and Costs

The first five Mercury research and development suits totaled $125,000. For the Apollo program, 60 suits were produced; and the contract cost of all functions, including manufacturing, was $90 million. This is the probable source of reports that the suits "cost" $1.5 million each. The first group of five Apollo Block II lunar suits bore a price tag of approximately $5 million. Hamilton Standard developed and produced, under separate contract, the portable life support system for $20 million. Each operational suit (not early research and development), depending whether it was IVA or EVA, cost between $90,000 and $125,000. Later Apollo suits, like the lunar or A7LB, had price tags ranging between $250,000 and $400,000. The major and most expensive problems remained the same: mobility, safety, and comfort for the user. It should be remembered that this cost included the large sums necessary for research and development of innovative technology—much of it concentrated on safety.

In a sense, the major cost factor has been safety: tight, thorough control of all parts and materials; complete inspection of all in-process parts at various stages of manufacture; extensive testing throughout the life of the suit; and continuous surveillance and control of the suit configuration and handling. More demands were made on the Apollo suit for each successive mission, requiring testing of every design change. Suit operations were labor-intensive and, therefore, expensive. The attention to quality and reliability apparently paid off: Not a single suit-related incident has caused a fatality, injury, or mission abort in any U.S. space program to date.[8]

● ●

adaptations of earlier air-cooled ventilated undergarments designed by Hans Mauch, a German scientist at the U.S. Aero Medical Laboratory.

The LCG resembled a pair of long johns, with an interlacing of capillary-like tubing intricately woven through nylon spandex attached to a comfort liner of nylon tricot. Water could circulate through this network of small tubes that rested practically on the astronaut's skin, cooling the body by direct conduction and almost eliminating perspiration. Warmed water would return to the PLSS, where it was cooled in a heat exchanger and circulated by a water pump. Integrated into the LCG were Nuweave crew socks with rib tops and flat knit feet. First worn by University of Pennsylvania oarsmen who wanted a sock that would stay up while they raced in stocking feet, crew socks became their generic name. This became the only style worn on the moon because the all-cotton construction did not generate static electricity in the spacecraft.[9]

The nylon spandex used for the liquid cooling garment consisted of 72% nylon polyester fibers knitted together with 28% elastic. Tricot for the innermost layer of the LCG was warp knitted from monofilament, dull or semi-dull nylon fibers. Tubing was lightweight, flexible, smooth, abrasion-resistant, transparent PVC plastic. Produced under the trade and extra thermal shielding between the lunar boot and the integral torso-limb suit boot. The outer layer of shielding, which acted as insulation, consisted of Teflon-coated Beta cloth, two layers of aluminized Kapton with integral Beta marquisette spacer, nine alternate layers of aluminized Mylar and Dacron scrim spacers, and two alternate layers of Kapton spaced with Beta felt. A layer of Chromel-R on the upper sidewalls of the boots protected against abrasion.[12]

Apollo crew members not assigned to work outside their spacecraft wore the five-layer torso-limb suit with a three-layer intravehicular cover layer (ICL) for added protection against fire. This was the outermost layer of the EMU. Astronauts assigned EVA tasks wore a 16-layer integrated thermal micrometeoroid garment over the basic PGA. The ITMG gave the same protection against fire, as well as added protection from extreme temperatures and from micrometeoroid impacts at speeds up to 30 kps (18 mps).

Several relatively new products were used experimentally as insulating material in the 16-layer ITMG. One was Mylar polyester film. Its tensile strength, resistance to chemicals and moisture, and ability to withstand temperatures from $-60°$ to $164°C$ ($-76°$ to $327°F$) made it a suitable candidate. Another excellent material incorporated into the Apollo ITMG was Kapton, first designated as an experimental "H" polyimide film, similar to Mylar at room temperature, but less affected than Mylar by temperature extremes. It can be combined with Teflon to produce a very stable material. Kapton is an insulating material that is exceptionally strong, thermally resistant, flexible, self-extinguishing, infusible, indissoluble, and resistant to high-energy radiation.[13]

The ITMG also had a Beta cloth outer layer. Scientists found that Beta cloth by itself creased easily and ripped. But Teflon added tensile strength and abrasion resistance when used as a coating over the Beta silica fiber, which was then spun into yarn (Figure 5.10). Under the Beta cloth were two layers of aluminized Kapton, integrated with Beta cloth marquisette—a light net-type material—used as a spacer to reduce heat conduction between Kapton layers. Five layers of aluminized Mylar spaced with nonwoven Dacron scrim came next. Nonwoven Dacron scrim came from a blend of polyesters heat-softened into webbing and fused mechanically into a bonded web, having the appearance of loosely woven material. The Kapton and Mylar sheets were perforated to bleed trapped gas and to prevent ballooning and rupturing during depressurization. The Kapton was found to wear, however, and was later replaced by Beta marquisette-reinforced polyimide film dubbed "super Kapton." The innermost layer of the ITMG was neoprene-coated nylon fabric, which served as a micrometeoroid bumper.

Intravehicular cover layers were made of two layers of woven Teflon-coated Beta silica fiber, completely fireproof even in a pure-oxygen atmosphere, and a Nomex* cloth inner layer, a heat resistant, non-melting polyimide material.

The astronauts' portable life support system, developed by Hamilton Standard, supplied oxygen, pressurized the space suit to at least 25.5 kPa (3.7 psi), and

---

*Nomex, a man-made fiber, was engineered and improved at this stage to be infusible. The original Nomex used in earlier suits had a melting point of $370°C$ ($700°F$).[14]

circulated oxygen through the helmet and suit. It controlled relative humidity and held temperature to a comfortable 21°C (70°F). A PLSS pump cooled water by sublimation* and recirculated it through the tubing of the liquid cooling garment. Carbon dioxide and other contaminants were filtered by a canister of lithium hydroxide and eliminated before they could accumulate and poison the astronauts' air supply. The 29-kg (65-lb.) PLSS contained enough water, oxygen, and other expendable supplies to last as long as four hours, depending on the rate of physical activity.[15]

The PLSS contained communication equipment that allowed lunar astronauts to carry on conversations with a fellow astronaut on the moon, with the pilot of the command module, or with Mission Control in Houston. This system also transmitted medical data on the crewman from his bio belt unit, which had sensing devices distributed to several body points. Physicians at Mission Control could monitor each astronaut and suggest adjustments in activity schedules when data indicated that the crew was being subjected to excessive work loads. All missions from *Apollo 7* through *Apollo 17* used this bioinstrumentation system.[16]

A two-part system was used for waste management. A urine collection container (Figure 5.11) was worn on the lower abdomen. Tubing connected to a transfer valve would allow deposit of the contents into the spacecraft reservoir. The solid containment system (Figure 5.12), for use

5.10
Doris Boisey and Delema Comegys from ILC work on Beta fiberglass marquisette, one of the high-technology materials used in space suit fabrication. Here it is spread out on long tables for finishing into spacers, materials used between layers of Kapton and aluminized Mylar for additional strength and insulation. These skilled seamstresses assembled the suits and sewed stitches to within tolerances of 2 mm (1/16 in.). (1967, SI photo 83-15829)

when the suit was pressurized, operated on the "diaper principle," absorbing and retaining moisture while oxygen washed odors downward away from the body. As spacecraft got larger, NASA engineers added on-board toilet systems for unsuited crew members. Solid waste was collected in bags containing a germicide to prevent gas and bacterial formation, then sealed and stored in the equipment bay.[17]

The Apollo suits were constantly modified as dictated by continuing research

---

*Hamilton Standard introduced a porous plate sublimator on the PLSS: Heated water from the LCG would pass through the sublimator, freeze at pores of the nickel plate exposed to ambient space temperatures, vaporize as heat was introduced through exchange fins, sublimate the ice film, and thereby free the vapor to be discharged.

● ●

Quality Control on Pressure Suits

As in the case of Mercury and Gemini, every astronaut selected for Apollo had three suits: one for flight use, one for training and simulation exercises, and one for backup. Pressure garment assemblies were classified based on the design purpose of each suit.

> **Class I—flight use (primary suit and backup); very tight control of fabrication, handling, testing, configuration.**
>
> **Class II—non-flight use, chamber; training, interface, certification testing; tight control of fabrication, testing, configuration.**
>
> **Class III—non-flight use, non-chamber; used for all other functions; no formal control of configuration.**

SOURCE: John Rayfield, ILC Dover, personal correspondence, January 2, 1987.

● ●

crometeoroid garments to ensure proper body temperatures and insulation from the extreme heat and cold of space. During his EVA, Schweickart was completely dependent upon his PLSS for oxygen, air-conditioning, and communications, unlike the Gemini astronauts who had been connected to their spacecraft with umbilicals.

The *Apollo 10* crew of Thomas Stafford, John Young, and Eugene Cernan proved the reliability of the lunar module two months later in May, and at last the painstaking work all seemed worthwhile. The United States was poised to send its first astronauts to the surface of the moon. *Apollo 11* crew members Neil Armstrong, Michael Collins, and Buzz Aldrin were launched on their historic journey on July 16, 1969.

Command module pilot Collins reminisced in *Carrying the Fire* about suiting up prior to launch. Once the crew donned their suits and helmets in the launch preparation area, they immediately began breathing pure oxygen to purge their circulatory systems of bubble-producing nitrogen. Carrying portable oxygen con-

tainers like suitcases, they then walked to the transfer van that transported them to the launch pad. The astronauts saw friends bid them a silent farewell, for all they were able to hear from inside their helmets was the hiss of oxygen coming into their suits (Figure 5.13).[20]

On prolonged flights, astronauts could now remove their space suits and don two-piece coveralls that afforded both comfort and protection. The coveralls, which were worn over a soft cotton undergarment, were made from Beta cloth, the same flame-resistant fabric as the space suit outer layer. The three astronauts did not remove their space suits and stow them under their couches until after the command and service module docked safely with the lunar module and they were on a proper trajectory for the moon. They wore their white inflight coveralls until it was time to suit up for the lunar landing.

Collins also donned a suit, albeit with the simpler three-layer ICL, even though he would remain in the orbiting command and service module. Armstrong and Aldrin, of course, donned suits complete

with the extra protective gear required for lunar exploration (Table 5.1). Finding mobility on the lunar surface easier than they had anticipated, they soon adapted a loping technique of one foot in front of the other. Both felt temperatures inside their suits were comfortable and that, overall, walking on the moon was a pleasurable experience.

Apollo 11 accomplished the goal set by President Kennedy: An American had walked on the moon and returned safely before the end of the decade. Six additional Apollo missions followed.

Apollo 12, with Charles Conrad as commander, Dick Gordon as command pilot, and Alan Bean as lunar module pi-lot, lifted off November 14, 1969. The crew was sent to collect more lunar samples and deploy the Apollo Lunar Surface Experiments Package (ALSEP), which contained a solar-wind spectrometer, ionosphere detector, nuclear generator, and magnetometer, and was designed to unveil the mysteries of the moon's internal structure. The crew would also examine *Surveyor 3* (an unmanned craft launched April 17, 1967, to survey the moon) and salvage laboratory samples for examination. Once the crew had returned to the Apollo spacecraft, they sent the LM crashing into the lunar surface, giving investigators an invaluable yardstick for reading seismic data.[21]

5.13
Apollo 8 astronauts Borman, Lovell, and Anders, wearing A7L suits, walk from the MSC building to the van which will take them to the launch pad. The astronauts are already connected to their portable oxygen supplies in preparation for launch. (NASA photo 68-H-1061)

**Table 5.1
Apollo A7L Materials
(See Figure 5.25)**

Layer*	Material	Function
	Extravehicular (EV) Suit	
	(rear entry zipper)	
1	Teflon cloth	Abrasion/flame resistance
2	Beta cloth (Teflon-coated silica fiber)	Fire protection (non-flammable in oxygen atmosphere)
3,5	Aluminized gridded Kapton	Reflective insulation
4,6	Beta marquisette (Teflon-coated silica fiber, laminated to Kapton)	Spacer between reflective surfaces
7,9,11,13,15	Aluminized Mylar	Reflective insulation
8,10,12,14	Nonwoven Dacron	Spacer
16	Neoprene-coated nylon	Inner liner
17	Nylon	Restraint layer for pressurized bladder
18	Neoprene-coated nylon	Bladder material serving as an impermeable layer containing suit-pressurization oxygen
19	Neoprene convolute	Pressure-retaining flexible joints
20	Knit jersey laminate	Abrasion protection
21	Lightweight Nomex cloth	Comfort
	Liquid Cooling Garment (LCG)	
22	Nylon spandex	Holds tubing close to skin
23	Vinyl tubing	Water distribution for cooling
24	Porous lightweight nylon	Comfort
	Intravehicular (IV) Suit	
1	Teflon-coated Beta cloth	Fire protection (non-flammable in oxygen atmosphere)
2	Nomex cloth	Snag/fire protection
3	Nylon	Restraint layer for pressurized bladder
4	Neoprene-coated nylon	Bladder material serving as an impermeable layer containing suit-pressurization oxygen
5	Lightweight Nomex cloth	Comfort
	Constant Wear Garment (CWG)	
1	Cotton	Comfort

*Materials are listed from outside to inside.

Lunar dust plagued the *Apollo 12* astronauts. Easily thrown up into clouds around them, it came down to cover their suits, tools, and scientific equipment. Problems with the thigh area of their suits caused the astronauts to have some uncertainty in balance. Lunar gravity caused them to lean severely forward in the loping motion. However, this did not prevent them from gathering surface samples.

Apollo 13 experienced more serious problems. An inflight systems failure jeopardized the on-board oxygen and power supply 56 hours after launch in April 1970. The lunar landing was immediately canceled, and the crew retreated to the lunar module, which they used as a lifeboat—still docked to the command and service module, for their return to earth. By doing so, they conserved the life support expendables of the command module for separation from the LM and reentry.[22]

There was no internal heat source to keep the cabin temperature at a comfortable level without command module power. The temperature stabilized at around 3°C (38°F), making the cabin so uncomfortably cold that the men had problems sleeping during their rest periods. Built to hold two people, the lunar module had insufficient room for three wearing bulky space suits. The Teflon inflight suits remained cold and clammy due to moisture build-up within the cabin. Two astronauts donned lunar boots for warmth, while the third wore an extra suit of underwear.

Had the *Apollo 13* crew carried out their mission as planned, Jim Lovell and Fred Haise would have been able to drink water from a new device dubbed "Gunga Din," a 237-ml (8-oz.) bag built into the space suit. The previous crew had complained of being thirsty while performing their tasks on the moon. Another new feature was a red band added above the elbows, below the knees, and on the helmet of one of the space suits to easily designate the commander. The helmet LEVA outer protective shell now contained provisions for "sun shades" on both sides and in the center, which could be adjusted to various heights.

Apollo 14, the third manned landing on the moon, touched down on February 3, 1971. To the astronauts' suits was added the buddy secondary life support system (BSLSS), established as a backup to the portable life support system. A long connecting hose, the BSLSS functioned in much the same way as emergency equipment for scuba divers when a companion's oxygen source failed. In case of PLSS failure, oxygen would still be supplied from the oxygen purge system (OPS), but cooling water would be cut off. In short walks from the lunar lander, the OPS oxygen flow would continue to remove body heat and keep temperature at acceptable levels. Longer EVAs demanded greater oxygen needs, which could possibly deplete the on-hand supply before the astronaut returned to his craft. Should it be needed, the BSLSS would extend the emergency oxygen supply between two astronauts 40 to 75 minutes.

The only major problem experienced with the space suits during this mission was some loss of mobility in Edgar Mitchell's right glove. Mitchell had just completed one of the longest lunar walks in an effort to explore the Fra Mauro area. He and Alan Shepard also had to drag or carry the modularized equipment transporter (MET) a good part of the distance because the dusty lunar soil impeded its mobility. It was also on this flight that Shepard, despite the stiffness of a confining space suit, managed to putt a record-breaking golf shot on the moon's surface.[23]

Examination of the plastic helmets worn by the crews of the first three lunar landings revealed cosmic ray damage. Cosmic rays were thought to account for reports by crew members of strange flashes of light, even when their eyes were closed. NASA specialists determined that astronauts on missions of several years' length would need extra shielding to prevent the destruction of nonregenerative body cells.[24]

In July 1971, *Apollo 15* astronauts extended their lunar extravehicular activity periods to twice as long as those of previous teams. This was made easier through ILC's redesign of the basic Apollo A7L suit, used since *Apollo 7,* to accommodate flexible convolutes instead of the built-in block-and-tackle mechanism that physically extended or contracted torso length. Improved convoluted joints were fitted at the knees, wrists, elbows, ankles, and thighs. Waist and neck flexion joints were added. They lessened fatigue when astronauts were suited up for long periods, as well as allowed mobility to bend over, touch the ground, and sit in the Lunar Rover. The new A7LB (first designated the A9L) suit contained the usual restraint-and-pressure bladder, but the long rear zipper now appeared on the right front, and a flexible container provided a greater supply of drinking water from inside the neckring.

Hamilton Standard also redesigned the PLSS increasing operation duration to seven hours. Oxygen tanks were thickened to withstand increased pressure and provide 50 percent more gas and 39 percent more water. Battery capacity was increased by 30 percent, and extra quantities of lithium hydroxide ensured longer life. Emergency OPS operations could be extended an additional 75 minutes. On earth, an entire EMU of this configuration with all consumables would weigh about 96 kg (212 lb.); this is equivalent to about 16 kg (35 lb.) on the moon.

Apollo 16 made its way to the moon in April 1972, and its crew experienced one mishap with their suits. The Gunga Din pouches inside the neckring of the PGAs leaked. By the time the two astronauts doffed their suits, sticky orange syrup (the fabled Tang) had practically immobilized their helmets until they poured enough water over them to release the locking mechanisms. Dirt was also found to be a potential hazard because it could become lodged inside the wrist connectors or embedded in the exterior surface of the space suit, where it could clog equipment.

The last of the manned lunar series, *Apollo 17* lifted off its launch pad in December 1972. The tasks assigned to mission commander Eugene Cernan included drilling into the lunar surface to collect sample materials. Cernan found pushing down on the drill difficult and painful. The extra effort needed to perform such physical activity while wearing a pressurized suit caused his heart rate to increase.

Before this last mission, suit technicians had added Velcro* pads in the helmets so that astronauts could scratch their

---

*Velcro is the brand name of the original hook-and-loop fastener. Velcro Fastening System's basic patent expired in 1978, and many competitors now produce similar fastening systems. But only Velcro makes Astro Velcro, which uses polyester hooks with Beta glass backing and Teflon loops to produce a fastener that will not burn.

noses while on the lunar surface, answering a complaint from earlier crews. Astronaut Ronald E. Evans tested a special orthostatic counter-measure garment. This device was designed to exert pressure to the wearer's lower limbs, thus preventing blood from moving too rapidly into the lower extremities during reentry.[25]

During the return trip, the command module pilots were tasked with retrieving cassettes and film magazines from the scientific equipment bay of the service module. An umbilical provided oxygen during these short space walks. An OPS identical to those worn during lunar EVAs acted as an emergency oxygen backup. The suits worn by the mission commanders and lunar module pilots retained the original six connectors, but they were more comfortably repositioned to form triangles on both right and left sides of the suits. The NASA emblem was also switched from the right side of the chest to the right shoulder on all suits.[26]

The successful flights of Apollo joined other great scientific voyages of discovery. People's perception of the moon would be changed forever by the historic missions and the knowledge garnered by the NASA teams. The trips to the moon fired the imaginations of millions of people, adults and children alike. Books on spaceflight, both fiction and nonfiction, proliferated in all languages, even for children (Figure 5.14). Science fiction films also enjoyed an enormous audience, who perhaps thought, in the mid-1970s, the possibilities of space discoveries knew no bounds.

Apollo suits reflected 34 years of developmental work. The managers of NASA's Manned Spacecraft Center believed they had provided the Apollo explorers with optimal solutions to a wide range of life

support problems through the latest state-of-the-art technology, materials, and processes. Despite the remarkable materials invented for them, Apollo spacecraft and space suits represented an expendable era. Most hardware was used only once and tailor-made for its mission. For future manned spaceflight missions, NASA sought increasingly practical, economical, and reusable hardware.

5.14

Fascination with space travel was worldwide. This illustration comes from the Russian children's book *Neznaika Na Lune (Neznaika on the Moon)*. (SI photo 84-11640)

5.15
The A1H (1964), with Mercury-type shoes, was rejected because its broad shoulders did not fit the Apollo couches. (NASA photo S-64-35127)

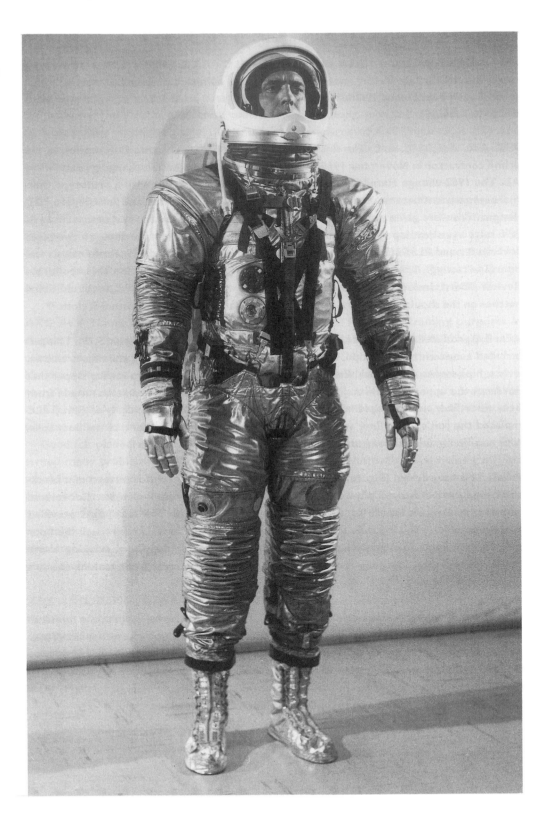

5.16
The A2H (1964) had Gemini-type shoes and aluminized neck convolutions. An MSC engineer *(left)* wears the prototype thermal overgarment, while a staff member models the prototype Apollo pressure suit. (NASA photo S-64-23432)

5.17 *(Left)*
The A2H Hamilton Standard/ILC pressurized state-of-the-art suit had aluminized covering, neckring, gloves, boots, and visor assembly. (NASA photo S-63-16605)

5.18
In the A3H prototype pressure suit, a white nylon cover layer replaces aluminum. Note that the white fiberglass helmet was still used for these human engineering and boilerplate tests, circa 1964. (NASA photo S-64-17091)

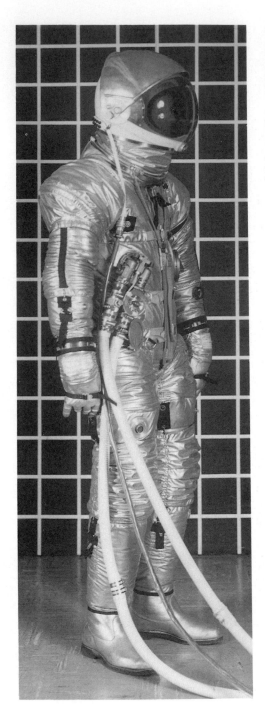

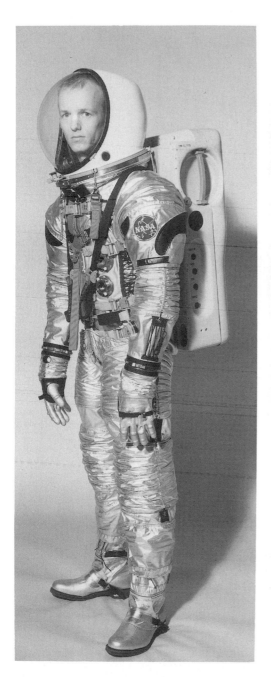

5.19 *(Left)*
The ILC state-of-the-art suit with an aluminized cover layer and PLSS became known as the A2L. A3L and A4L designations were given to obsolete suits such as this on which design changes were tested. (NASA photo S-65-17251)

5.20
Dr. Joseph Kerwin models the A4H prototype at MSC, October 1965. The designs of the helmet and the neckring are new. (NASA photo S-65-54961)

5.21 *(Left)*
ILC won NASA's space suit competition in 1964 with their AX5L suit. (NASA photo S-66-33934)

5.22
A3L and A4L were designations given to now obsolete suits used to test design changes, 1964. (SI photo 87-6802)

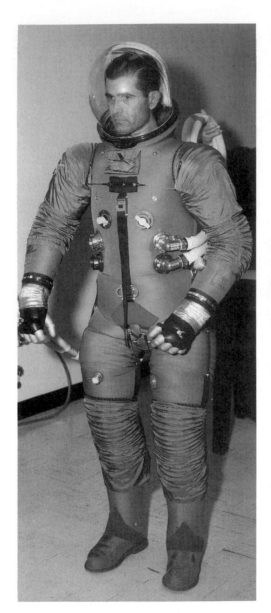

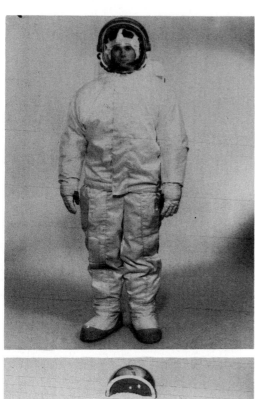

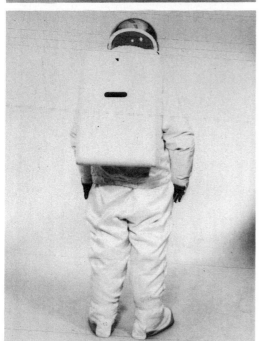

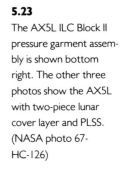

5.23
The AX5L ILC Block II pressure garment assembly is shown bottom right. The other three photos show the AX5L with two-piece lunar cover layer and PLSS. (NASA photo 67-HC-126)

5.24 *(Left)*
The A6L suit (1967) adopted design changes recommended by the Apollo 204 Review Board, including Beta fabric in the cover layer and fire-resistant materials wherever possible. Note the Chromel-R patches at elbows and knees. (NASA photo 67-H-1116)

5.25
This basic A7L pressure garment assembly has the integrated thermal micrometeoroid garment removed. The engineering challenge was the design of the joints. At joints, a handwoven linknet caused restraint in all directions of the accordion-like convolutions. Steel cables connected the shoulder joint to plates on the back and chest and limited knee flexion to the vertical fore-and-aft plane. This suit was the product of 18 years of research, cost more than $100,000 to produce, and was the most complex tailored outfit assembled at that time. (NASA photo S-71-24537)

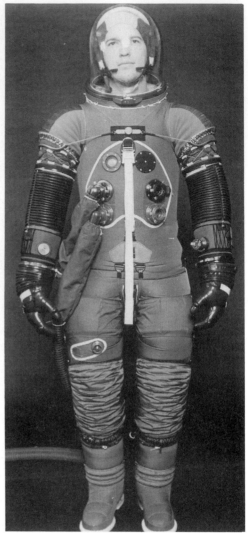

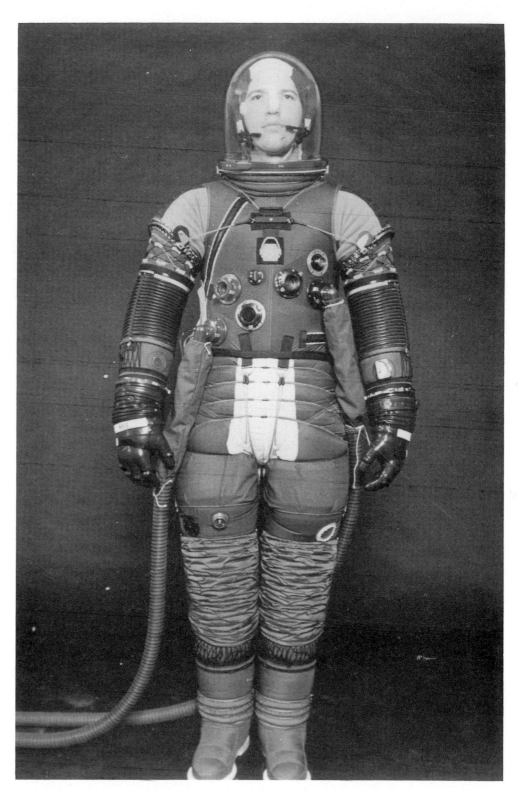

5.26
The A7LB suit from ILC
(1971) is shown without
the ITMG. Note the im-
proved convoluted joints
at the knees, wrists, el-
bows, ankles, and thighs.
(NASA photo
S-71-24533)

Table 5.2
Apollo A7LB Materials
(improved lunar suit used on *Apollo 14, 15, 16,* and *17*)
(See Figure 5.26)

Layer*	Material	Function
	Extravehicular Suit	
	ITMG Cover Layer	
	(frontal zipper)	
1	Teflon cloth	Abrasion/flame resistance
2	Beta cloth (Teflon-coated silica fiber)	Fire protection
3,5,7	Beta marquisette (Teflon-coated silica fiber)	Spacer between reflective surfaces
4,6	Aluminized gridded Kapton	Fire/thermal radiation protection
8,10,12,14,16	Aluminized Mylar perforated film	Reflective insulation
9,11,13,15,17	Nonwoven Dacron	Spacer
18	Neoprene-coated nylon	Liner (ripstop)
	Basic Torso-Limb Pressure Garment Assembly	
19	Nylon fabric	Restraint layer for pressurized bladder
20	Neoprene-coated nylon	Pressure-retaining flexible joints (convolute/bladder)
21	Lycra jersey	Abrasion protection
22	Lightweight Nomex cloth	Comfort; air diffusion; aid in donning PGA
	Liquid Cooling Garment	
23	Nylon spandex	Holds vinyl tubing close to skin
24	Vinyl tubing	Water distribution for cooling
25	Porous lightweight nylon	Comfort
	Intravehicular Suit	
	Cover Layer	
1	Beta cloth	Fire protection
2	Nomex cloth	Snag/fire protection
3	Beta cloth	Fire protection

*Materials are listed from outside to inside. Layer numbers are representative; layers change according to mission needs.

CHAPTER 6

Space Suits
for the
Manned Orbiting Laboratory,
Skylab,
and the
Apollo-Soyuz Test Project

Planning for post-Apollo projects began in 1964 at President Johnson's request. The Office of Manned Spaceflight wanted a manned mission to Mars. The Office of Space Science, including Wernher von Braun, sought a manned orbiting laboratory (MOL) or a space station with reusable shuttles. Because costs of new programs were unpredictable and could be considered excessive, NASA director James Webb chose a soft-line course. Annual spending would run between $4 billion and $5.7 billion. By 1967, the Space Science Board called for a deemphasis on manned spaceflight. Budgetary pressures, inflation, war in Vietnam, and environmental concerns overshadowed the space program. In 1970, President Nixon deferred plans for the space station until we had a working shuttle. *Apollo 18* and *19* missions were scrubbed. The Applications program was reduced to a single Skylab. Nixon canceled the MOL program. But development continued despite these limitations.

MANNED ORBITING LABORATORY

During the 1960s, advanced planners at both NASA and the U.S. Air Force were studying the desirability of a permanent earth-orbiting space station. Such a station could serve as a staging area for expeditions beyond the vicinity of earth and as a vantage point for earth-directed activities. It is interesting, in light of the current struggle to achieve a viable long-term space station, to think how far back the concept goes.

In mid-1963, NASA had proposed a manned orbiting laboratory as a post-Apollo project. This could be either a small, low-cost laboratory taking advantage of hardware already developed or a larger structure that could support four to six people for a year. In either case, astronauts could conduct a variety of scientific experiments, including biomedical and astronomical tests. However, NASA's studies did not proceed beyond the design stage.

The Air Force's program for manned spaceflight, involving a plane known as Dyna-Soar (X-20), was canceled in late 1963. Funds for the X-20 were redirected to an Air Force Manned Orbiting Laboratory that would combine NASA Gemini-era spacecraft and launch vehicles with newly developed hardware. MOL was to be an operational facility designed for extravehicular activity, laboratory operations, and long-term stays. Schedules called for the first mission to take place in 1967–68.[1]

The Air Force sought a new lightweight space suit, possibly for continuous wear, to be developed in support of MOL. Mercury and Gemini suits were obsolete, and Apollo suits were being designed specifically for the lunar program. The service, therefore, promoted the study of advanced suit design at ILC, Hamilton Standard, and Litton Industries. Litton was experimenting with a totally new concept for a suit made from hard materials.

Space suit designers sought to diminish the negative effects of long-term weightlessness and associated environmental factors. Problems envisioned for MOL crews included loss of muscle tone and circulatory difficulties. Studies indicated the effectiveness of exercise combined with the use of pressure cuffs (something like those used to measure blood pressure) or breathing techniques to alleviate these problems.

Gemini EVA experience revealed the toll extracted on the astronaut. Fatigue was a real problem and astronauts suffered calcium loss from bones and nitrogen loss from muscles, severe enough to pose potential danger if endured for long periods. No studies had been done in Apollo because the situation had not been considered serious on short-term flights. Several studies were planned for the orbiting laboratory. Diet would be an important control for proper measurement of physiological losses. Mineral balance studies, however, are exacting procedures even in well-equipped hospitals. Accurate measurement aboard the spacecraft would be difficult.[2]

Notwithstanding their concern for an advanced space suit design, Air Force experts studied one- and two-gas atmospheres for the MOL that would not require continuous use of a suit. In an environment with a temperature of 29°C (85°F) and a nitrogen-oxygen atmosphere at 34 kPa (5 psi), astronauts could work comfortably wearing cotton summer flying suits and underwear.

As prime contractor for the Air Force MOL suit since 1967, Hamilton Standard had produced a well-designed suit (Figure 6.1) with demonstrated ease of mobility, donning, and doffing. But the contract was canceled in July 1969. In June, the Department of Defense had canceled the MOL program in response to calls for budget cuts by Congress and to advances in unmanned satellite data-gathering systems. The seven astronauts in training for MOL transferred to NASA's astronaut corps.

Although considered briefly for Apollo, the MOL suits were in the early stages of production and could not be converted for use by NASA at that late date.[3] Several MOL prototype suits are in the National Air and Space Museum's preservation collection at the Garber Preservation, Restoration, and Storage Facility.

SKYLAB

On July 22, 1969, just two days after *Apollo 11* landed on the moon, NASA received directions to implement an Advanced Apollo Program, which would include an orbiting workshop. The name Skylab, for "laboratory in the sky," was chosen to overcome dissatisfaction with the abbreviation AAP, which congressional opponents jokingly came to refer to as "Almost a Program."[4]

The workshop itself would be crafted from a Saturn launch vehicle stage and orbited by the Saturn V rockets. In early manned missions, astronauts gained access to the vacuum of space by first depressurizing the relatively small cabins of their Gemini or Apollo spacecraft and evacuating their life-supporting atmosphere. Depressurizing and repressurizing

the huge interior of Skylab's workshop was out of the question. Instead, redundant Gemini hatches were fitted to the circular wall to form an air lock module where two suited astronauts could comfortably wait. Thus only 5 percent of the Skylab environment was depressurized for EVA. Skylab would also use a nitrogen-oxygen atmosphere. Physicians worried that long durations in pure oxygen would cause permanent damage to the astronauts.

Skylab was a totally new experience for design engineers. Manned earth orbit operations would carry astronauts to the highest inclinations yet. Three-man crews would visit Skylab via Apollo spacecraft,

6.1
This Hamilton Standard suit developed for earth orbital missions was intended for intravehicular and extravehicular operations. Although it had excellent mobility, the suit never reached full development. NASA chose the Apollo ILC suit which was already in its final stages of development with a functioning portable life support system. (SI photo 87-15043)

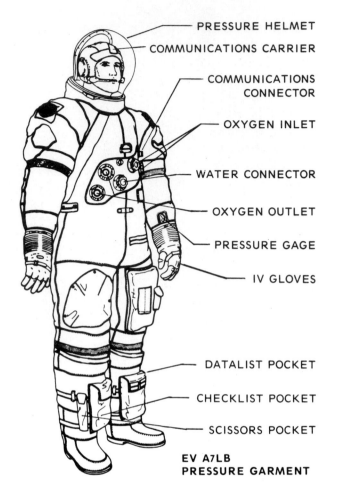

PRESSURE HELMET
COMMUNICATIONS CARRIER
COMMUNICATIONS CONNECTOR
OXYGEN INLET
WATER CONNECTOR
OXYGEN OUTLET
PRESSURE GAGE
IV GLOVES
DATALIST POCKET
CHECKLIST POCKET
SCISSORS POCKET

**EV A7LB
PRESSURE GARMENT**

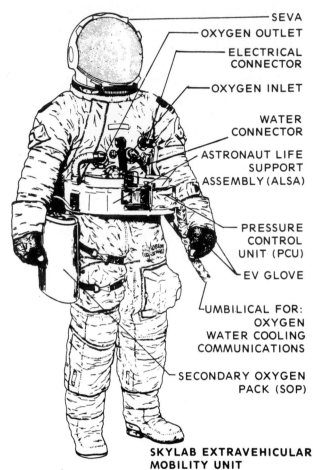

SEVA
OXYGEN OUTLET
ELECTRICAL CONNECTOR
OXYGEN INLET
WATER CONNECTOR
ASTRONAUT LIFE SUPPORT ASSEMBLY (ALSA)
PRESSURE CONTROL UNIT (PCU)
EV GLOVE
UMBILICAL FOR:
OXYGEN
WATER COOLING
COMMUNICATIONS
SECONDARY OXYGEN PACK (SOP)

**SKYLAB EXTRAVEHICULAR
MOBILITY UNIT**

which would be launched separately. There would be EVAs on each mission. This program allowed NASA to take advantage of spacecraft and launch vehicles already procured for Apollo but no longer required with the cancellation of missions to follow *Apollo 17* (1972). The Skylab orbital cluster was launched in May 1973. Three manned Skylab missions were conducted during the next months.

Skylab astronauts wore modified Apollo A7LB pressure suits during launch and

extravehicular activity. The Skylab pressure garment assembly for EVA (Figure 6.2) interfaced with the spacecraft environmental control system through an umbilical connected to the astronaut life support assembly (ALSA) worn on the astronaut's chest. Since long space walks were not necessary, the ALSA, rather than an independent PLSS, regulated air pressure at 26 kPa (3.7 psi), as well as water, oxygen, and electrical power, and was designed to withstand temperatures from

−117° to 136°C (−180° to 277°F). The ALSA was somewhat similar to Gemini's extravehicular life support system (ELSS), except that Gemini suits held pressure at 34 kPa (5 psi) and were air-cooled, thus not needing circulating water.

The Skylab suit included mobility joints at the knees, wrists, shoulders, elbows, ankles, and thighs to ensure normal body flexure. Mobility joints at the neck and waist reduced fatigue, conserving the wearer's energy. Entrance zippers extended around the back from the left side of the waist to the right side and then diagonally up to the right chest area to accommodate the new neck and waist joints.

In addition to a new lightweight integrated thermal micrometeoroid garment covering the PGA, astronauts wore the Skylab extravehicular visor assembly (SEVA), which included a shell over the helmet. There were two visors, two side eyeshades, and a center eyeshade mounted in the outer shell. The sun visor, thinly coated with gold, reflected solar heat and light. The inner visor, which by contrast was coated to retain heat, was transparent. This visor was also used without the sun visor in areas of shadow. The visors and eyeshades could be adjusted for comfort and safety, as demanded by dark-side and sun-side conditions.

A new ventilation system removed body heat during EVA. The liquid cooling garment with its network of polyvinyl tubing circulated cool water from the portable life support system over the body, transferring body heat to the liquid and then rerouting the liquid back through the PLSS. This garment was used for an unexpected purpose during the first Skylab

mission when the crew failed to control a serious heat loss. Ground control suggested attaching the LCG to the long umbilical. The LCG attachment, placed near the water tanks, then provided a path for heat to reach the multiple docking adapter's cooling system, preventing the coolant loops from freezing.

The Skylab EVA boots were slip-on assemblies made from materials similar to those in the ITMG. There were two pairs of gloves: intravehicular gloves with a single-wall restraint and bladder structure formed to fit the astronaut's hand and outer thermal gloves made of the same materials as the ITMG. The outer shell of the glove, constructed of metal-woven fabric with silicone rubber-capped fingertips, provided thermal and abrasion protection.[5]

Skylab crews became the first to wear inflight coverall garments rather than EVA suits for reentry and landing. Crew members wore brown inflight assemblies (serving the function of jacket, trousers, shirt, boots, and gloves) fabricated from fire-resistant woven Durette fabric. Trousers could be converted to shorts with the pull of a zipper. Since there was no provision for laundering clothes on board Skylab, outfits were worn for a few days and then discarded. Technicians had packed enough clothes for each astronaut for the duration of the mission.

Each clothing module contained 14 sets of underwear.* Each set included four different types of underwear, all made of cotton: a full union suit with integrated

6.2 (Facing page) Modification to the A7LB pressure garment included an astronaut life support assembly (ALSA) worn on the chest rather than on the back as the PLSS was. The ALSA regulated water, oxygen, and electrical power received from Skylab through an umbilical. An emergency supply called the secondary oxygen pack (SOP) was carried on the left hip. Unlike Apollo missions, Skylab was to be a routine operation to find out how well astronauts could work at regular tasks aboard and around an orbiting space station. Skylab reflected the shift from exploratory, deep-space operations to earth-orientated activities. (NASA photo 72-H-1161)

---

*During exercise, astronauts stripped to their underwear.

6.3

Skylab inflight boots, or portable foot restraints, incorporated cleats on the soles to anchor the astronaut in place on the Skylab's grid floor for activities that required stability such as meals. (SI photo 90-8202)

socks; a half union suit (bottom) with integrated socks and a standard pullover T-shirt; jockey, knee, or boxer shorts and socks; and an Apollo constant wear garment. Skylab inflight boots, called portable foot restraints (PFR), incorporated cleats on the soles allowing crew members to anchor themselves on Skylab's metal grid floors (Figure 6.3). This way they could remain relatively stable during work periods. Astronauts, however, found that bending over in zero gravity to pull on socks or tie shoelaces was difficult.[6]

Mealtimes were welcome breaks. Each morning the crew assembled around the wardroom table for breakfast. The astronauts preferred to stand because sitting also placed a strain on stomach muscles from the forced bending at the waist. Food had improved since Apollo days but the astronauts still complained that it was too bland.

Skylab's waste management compartment resembled a commercial jetliner's bathroom in size, metallic appearance, and gurgling noises, but it took some getting used to. Engineers had given the floor a smooth surface for easier cleaning, making it difficult to gain a foothold. Astronauts also complained that toilet articles floated away unless anchored. Hands were washed from a valve recessed in the wall. The first crew showered about once a week and did not seem to mind vacuuming up excess water. Other crews settled for daily washcloth scrubs since a shower took almost an hour.

SKYLAB MISSIONS

The final mission of Saturn V lofted *Skylab 1* into its planned orbit on May 14, 1973. The crew of *Skylab 2,* the first to visit the orbiting workshop, remained at the lab for 28 days. Following a flawless lift-off on May 25, 1973, the crew faced an extraordinarily demanding assignment. Skylab had entered orbit with damaged solar wings and without the protection of the heat shield, which had been damaged during its spaceflight. Sunlight beat mercilessly on the space station raising the temperature and making the station uninhabitable. Food, medicine, and other supplies might all have been ruined. Engineers at Houston's control center studied the problem and tested possible solutions. A team of specialists, including seamstresses Delores Zeroles (Figure 5.7) and Ceal Webb, flew in from ILC Industries to Huntsville to work on emergency sunshade designs.

During the first stand-up EVA attempt at repairs, Paul Weitz manipulated the tools while Joseph Kerwin held his legs and Charles Conrad maneuvered the spacecraft. Their job was to dislodge debris fouling the solar array boom deployment. Aluminum strapping tightly secured the solar array boom. Weitz pulled with all of his strength but to no avail. Even docking attempts failed. The probe would not engage the drogue. The astronauts depressurized the spacecraft, opened the forward tunnel hatch, and removed the probe's back plate to bypass electrical connections. They centered the probe and drogue and used Apollo's thrusters to close on the docking adapter. After the docking surfaces met, the latches properly engaged. They were successfully hard docked.

On their second day in space, the astronauts ventured another EVA, during which Conrad and Weitz almost fully extended a parasol over the workshop, which brought the temperature within a bearable range. The crew, in coordination with ground control, carefully planned for their EVA boom deployment task. This would take the crew into an area nobody expected, so no handholds or foot restraints had been provided.

Conrad and Kerwin drifted outside. They assembled a long pole and affixed the metal cutter. As Skylab orbited the earth, the crew worked through each daylight period and waited out each nightfall until they could continue their risky spacewalks. Conrad and Kerwin, still tethered to the spacecraft, pulled hard on the cutter jaws balanced over the metal strap wrapped around the meteoroid shield but nothing happened. Just as Conrad moved back up to the cutter to check it, the jaws worked, snapping the strap apart, sending the boom away from the spacecraft and Conrad cartwheeling into space. He managed to grapple back onto the pole with his tether. Now, he and Kerwin had to force the boom into a 90-degree rotation. Standing on the workshop wall, Conrad pulled one end of a rope upward over his shoulder while Kerwin pulled from the opposite end. The bracket snapped and the boom moved out to the full 90 degrees. Skylab would have nearly twice the power it had struggled with to this point.

Another scored point on EVA: The crew's fully extended solar arrays of the telescope mount provided needed kilowatts of power. By mission's end, this crew had met nearly 100 percent of the medical requirements; 80 percent of the solar observations; and 60 percent of the earth resources experiments, due in great part to their successful EVAs.[7]

Skylab 3 astronauts logged 59 days in July and August 1973. Owen Garriott and Jack Lousma carried out a six-and-a-half-hour EVA during which they deployed a new sunshade over the old one. They concluded their EVA by retrieving experiment samples, changing film in the Apollo Telescope Mount (ATM) telescopes, and examining Skylab's exterior for damage. For their second EVA, Alan Bean stayed in the command module while Garriott and Lousma went outside to change film and plug in a new gyroscope package. When all of their tasks were finished, Lousma reluctantly returned to the spacecraft. He said it was an exhilarating experience floating freely more than 400 km (250 mi.) above earth.

The third and last crew, *Skylab 4,* remained 84 days during November 1973 through January 1974. When Gerald Carr, Edward Gibson, and William Pogue entered Skylab, they were greeted by a message on the teleprinter and three stowaways. On the bicycle ergometer and the lower-body-negative-pressure device and comfortably poised on the toilet seat sat three stuffed space suits looking very much like crew members.

As Gibson and Pogue prepared for their first EVA, they found mildew on the liquid cooling garments. Ground control advised them to wipe them clean and hang them to dry in a warm part of the workshop. On their second EVA, they repaired antennae, installed ATM film, pinned open a sticking door, deployed various experiments, and photographed the earth's horizon at sunset for a coronagraph contamination experiment.

EVAs never being routine, Carr noticed a large build-up of ice on his chest pack that regulated oxygen and water flow. The pressure control unit was critical to the life support system because it was designed to accept consumables from the space station and feed them into the space suit. Carr had exchanged pressure control units with Pogue, forcing an O-ring seal on the water connector which caused liquid to leak and freeze on the Beta cloth cover of his space suit. Fortunately, the frozen mass had not built up enough to jam the chest pack control valves for this could have had disastrous results.

All three crews performed critical extravehicular activity as they repaired and maintained Skylab and tended scientific instruments mounted on the workshop's exterior. The modified Apollo suits, having stood up to nearly 42 hours of EVA, met the needs of the Skylab program. Several new maneuvering techniques, including the astronaut maneuvering unit (Figure 6.4), were evaluated aboard Skylab.

APOLLO-SOYUZ TEST PROJECT

The United States and the USSR had been competitors in space since the late 1950s, first with unmanned satellites and then with manned spacecraft. Many people felt the "space race" culminated with the first lunar landing. But acting NASA Administrator Thomas O. Paine believed that the time was right for these longtime competitors to pursue cooperative ventures.

The first exploratory discussions between technical representatives of the two countries took place in 1970. A project to demonstrate this new avenue of cooperation had been agreed upon by 1972. An American Apollo and a Soviet Soyuz, launched separately, would dock in space. Connected by a unique, jointly designed

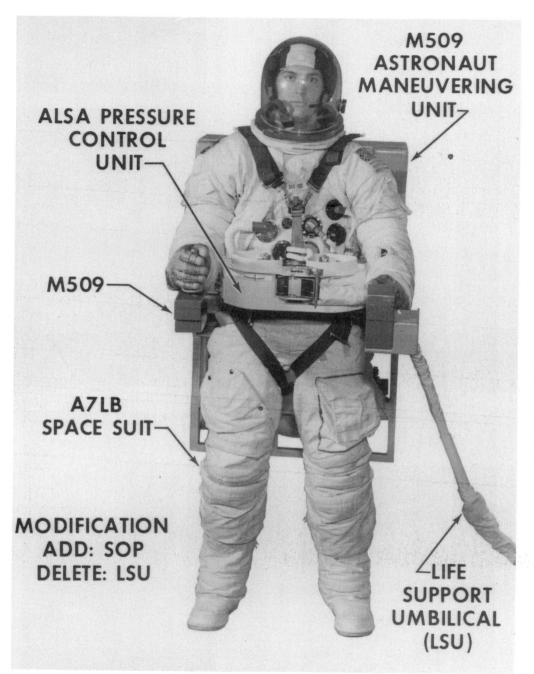

M509
ASTRONAUT
MANEUVERING
UNIT

ALSA PRESSURE
CONTROL
UNIT

M509

A7LB
SPACE SUIT

MODIFICATION
ADD: SOP
DELETE: LSU

LIFE
SUPPORT
UMBILICAL
(LSU)

6.4
During *SL-3*, Alan Bean was the first astronaut to fly around the spacious workshop dome using the automatically stabilized astronaut maneuvering unit (AMU) built by Martin Marietta and developed from Gemini days. The AMU was equipped with tanks of nitrogen to prevent contamination of the Skylab atmosphere and had 14 thrusters to control attitude. Successful tests by Owen Garriott and Jack Lousma as well proved the device would allow space-walking astronauts to move from one vehicle to another or to station themselves at one location for work or inspection purposes. (NASA photo S-70-5178-S)

6.5 *(Top)*
This photo of Skylab shows the sunshade correctly in place. When the crew of *Skylab 2* arrived at the space station, they found the heat shield damaged. A team of experts gathered in Houston to work with the crew on installing a sunshading parasol. The workers on earth clapped and cheered when astronauts Conrad and Weitz extended the parasol over the workshop. (SI photo)

6.6
The 1975 ASTP crew: American astronauts Donald "Deke" Slayton, Thomas Stafford, and Vance Brand with Soviet cosmonauts Aleksey Leonov and Valeriy Kubasov. (SI photo 87-7734)

● ●

Political Impacts

During the height of the Apollo heyday, the Kennedy Space Center had employed more than 26,000 people. By 1974, tremendous changes had taken place. At this time, there were only about 10,000 employees. The Cape's once-swollen budget of $500 million had dropped to $219 million. NASA's overall budget was reduced to one-third of what it had been ten years before. David Baker suggests in *The History of Manned Spaceflight* that the American government lacked vision and commitment, responding too quickly to wavering public moods. In Baker's opinion, the United States is the most effectively equipped machine to accomplish massive technical and industrial programs, but, he believes, the drive to realize such dreams dies under commercial interests. Thus, money for space exploration remained scarce for the next several decades.

● ●

docking module, the two crews would visit one another in orbit, performing symbolic, scientific, and applications activities. The Apollo-Soyuz Test Project (ASTP) took place in July 1975 (Figure 6.6).[8]

The Soviets used a sea-level atmosphere in their spacecraft, while the Apollo was designed to support a pure-oxygen atmosphere. To provide an air lock between vehicles, almost identical docking units were constructed for each spacecraft, differing only in the method of docking. The Apollo Docking Module contained normal sea-level atmosphere on the launch pad. As the spacecraft ascended, it slowly vented the nitrogen until it reached one-third sea-level pressure. Before opening up the tunnel that linked the Apollo and Soyuz, astronauts evacuated the nitrogen and pumped up with oxygen to one-third sea-level pressure. Once inside the module, they closed the hatch and added nitrogen, bringing up the pressure to two-thirds sea-level and making it compatible with Soyuz. The Soyuz hatch was

then opened and the astronauts moved into the Soviet spacecraft.

Since extravehicular activity had been excluded from this joint mission, the three astronauts needed space suits only for launch, orbital insertion, docking and undocking exercises, and the return trip. NASA chose a sophisticated modification of the A7L model suit with an intravehicular cover layer, but gave it an A7LB designation (Figure 6.7).* The only problem experienced with this model arose during a countdown demonstration test when one suit leaked. Technicians quickly isolated the problem and made minor modifications to the pressure-sealing slide fasteners on all three suits.

The two cosmonauts wore soft pressure suits with soft built-in helmets (Figure 6.8), rigidly fixed, with zippered visor

---

*For the ASTP, suit entry/closure zippers ran down the back and up through the crotch, whereas the A7LB's zipper ran down the right front side. The A7L and A7LB suits had six connectors while the ASTP suit had only three.

6.7

For the Apollo-Soyuz
Test Project, NASA mod
ified the A7L with an in-
travehicular cover layer
and designated it A7LB.
Here crewman Vance
Brand is suited up for a
July 1975 countdown
demonstration test.
(NASA photo 75-H-738)

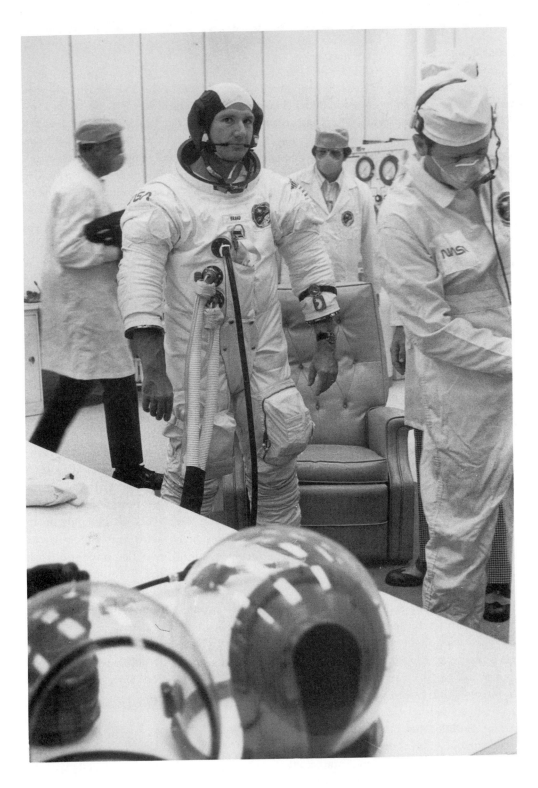

closings and removable gloves. A ventilated undergarment, which relied on a system of sewn-on tubes to transport gas from its own supply system over the wearer's body to maintain body temperature, completed the suit assembly. The pressure suit was made of four layers: two airtight layers, one liner, and one safety layer. The outermost layer acted as a heat shield. Suit mobility was a product of the elasticity of the material and the gas pressure within the garment.

For most inflight activities, the Americans wore coverall garments similar to those worn by Skylab crews. New to the ensemble were soft ventilated shoes with Velcro fasteners and soles.

The cosmonauts also wore coverall garments, but a material had to be found that was acceptable to Apollo's nearly pure-oxygen atmosphere. Because all materials in both spacecraft had to be non-

6.8
Soviet ASTP suits are displayed at the Cosmonaut Training Center at Star City near Moscow. (NASA photo 75-HC-238)

● ●

Politics and Progress

Despite ASTP's accomplishments, the international cooperation in space, science, and technology soon came to a halt. By 1976, Soviet deployment of new generations of strategic military missiles created political tension, and talks between the two nations dwindled down to nothing. On December 28, 1979, Soviet infantry entered Afghanistan, beginning a war that alienated the U.S. and other Western nations and ended U.S.-Soviet cooperation on the space program.

In *The History of Manned Spaceflight,* David Baker claims it was the bold pride and commitment of a New World people that met the challenge of Apollo. Manned spaceflight might have died in the 1960s had not President Kennedy committed the United States to the job. For now, the legacy of Apollo rides upon the wings of the Shuttle.

● ●

flammable, the Soviets developed a fireproof material called Lola. It was found to be superior to the American material in some ways. Lola cloth had self-extinguishing characteristics; the American cloth would burn, although very slowly. Each cosmonaut also had access to a wool cardigan to be worn under the jacket if desired. The constant wear garment, part of the inflight clothing, was made of cotton/linen knit. The cosmonauts' light leather boots were protected by covers made of Lola cloth.⁹

The two crews completed their joint and individual agendas successfully during the nine-day mission.* This was the last time the Apollo spacecraft and Apollo suit would see service. On the reusable Space Shuttle, the next U.S. crews would wear a new generation of space clothing.

———

*The only danger to the U.S. crew occurred during the descent phase. The cabin accidentally filled with noxious fumes, residual fuel, and oxidizer, causing the astronauts to cough and feel nauseated. Splashdown was hard and the command module rolled over suspending the three men upside down. The crew began breathing pure oxygen from face masks. Thomas Stafford was able to activate the uprighting system and open valves that dissipated the toxic fumes.

CHAPTER 7

Space
Shuttle Suits

Details of the revolutionary concept of a reusable manned space transportation system were made public in 1972, when the Nixon administration announced a new program intended to make earth-orbital travel almost routine. A winged orbiter, designed to be capable of at least 100 missions, would be launched by its own three rocket engines. Two solid-fuel rocket boosters would supplement the launch of the 91,000-kg (100-ton) orbiter. The Space Shuttle orbiter crew would perform services such as satellite deployment, satellite rescue missions, and repair in low earth orbit, then glide the Shuttle back into the atmosphere unpowered, landing like an airplane. The craft would be ready for another mission after servicing and refurbishment.

Scientists thought it possible to tie Shuttle missions into reuse of the Skylab's facilities. Skylab's three long-duration missions proved beyond doubt that humans could perform scientific work while orbiting the earth. Analysis of medical data from these missions showed humans adapted well to zero gravity, and that given proper environmental controls, they could maintain physical well-being and morale and then readapt to earth conditions. By 1977, NASA was planning further Skylab missions.

The scientists soon had doubts about how much longer Skylab could stay in orbit. It was unable to maintain a position in sunlight long enough to recharge fully. Its systems demanded ten-hour shifts manned by many different NASA crews. In July 1978, NASA set up a panel to determine how to deal with Skylab's reentry into earth's atmosphere. By December, schedules for Shuttle program development had slipped badly and Skylab's orbit was rapidly decaying. It was all over July 1, 1979: Skylab showered Australia with spectacular fireworks as the spacecraft burned up on reentry into earth's atmosphere.

Dreams of a permanent American space station were temporarily scuttled and NASA went back to grappling with Shuttle problems. *Columbia,* the first Shuttle orbiter, made its maiden flight in April 1983, followed by the orbiters *Discovery* and *Challenger.*

NASA planned several major changes in space apparel for manned Shuttle flights. A selection board evaluated two proposals for Shuttle suits: one from Hamilton Standard teamed with ILC, the other from AiResearch Manufacturing Company teamed with David Clark. In 1976, NASA once again selected Hamilton Standard as the prime contractor to oversee development of the Shuttle space suits and to manufacture the critical primary life support system (PLSS).* ILC Industries, Inc.† was responsible for the pressure suit under subcontract to Hamilton Standard.[1]

The time had come to progress beyond single-mission space suits. Reusable space suits for the Shuttle program incorporated improvements generated from an advanced hard suit technology project that provided multisize equipment with lifetimes of eight years for soft goods (textiles) and 15 years for hardware. Shuttle suits were to be manufactured as modular components in small, medium, and large sizes. Extra-small and extra-large sizes were added in January 1978.

The Crew Systems Division of NASA's Johnson Space Center decided in the mid-1970s to use flight-ready suits for certification and validation testing, rather than rely on engineering or prototype models. It was also decided that crew members would not don space suits on routine Shuttle missions unless they were to perform some extravehicular task in the vicinity of the Shuttle cargo bay.

Gimbaled scye bearings‡ in the hard upper torso (HUT) made movement and putting the suit on easier. Changing HUT construction from aluminum to fiberglass led to improved manufacturing methods. Fiberglass cut overall suit weight and was easier to machine and shape. Adding secondary axial restraints to the limbs increased safety. Suited astronauts had more stability.

The high-pressure oxygen regulator required redesign after a flash fire destroyed an empty suit during routine tests. The resulting component material changes cost approximately $11 million but

---

*Previously, PLSS referred to the *portable* life support system. Because a secondary system was employed, PLSS refers to *primary* life support system in the Shuttle program.

†Through various mergers, resales, and acquisitions, International Latex Corporation's ILC Government and Industrial Division became known simply as ILC Industries, Inc.

‡*Scye* is a tailoring term for the shape or outline of an armhole. A *gimbal* is a device consisting of two rings mounted on axes at right angles so that an object will remain suspended in a horizontal plane between them regardless of any motion of its support, as in a ship's compass.

greatly improved the suit. The ventilator fan motor was redesigned following a failure to operate properly during the first intended use of the extravehicular mobility unit (EMU), on the STS-5 flight in November 1982. Added field test equipment improved the efficiency of refurbishing and checking EMU hardware between flights because testing could be done at NASA rather than at the factory. The program required constant replanning to reflect the continual, incremental slippage in Shuttle flight dates. All these changes added to costs. An initial estimate of $46.9 million to cover 43 sets of suit components and 13 life support systems grew to an actual cost of $167 million through the first EMU use in space, during the STS-6 flight in April 1983. NASA expected savings would result from repeated wearings of the suits.[2]

Shuttle suits* are assembled from modular parts manufactured in various sizes: five standard sizes of torso, a variety of sleeve lengths and lower torso sizes, and a "one size fits all" helmet. The suit can be broken down following each mission so parts can be cleaned, dried, and then reused by other astronauts on subsequent missions. Since 1985, only gloves (Figure 7.1) have been custom-fitted to each astronaut to relieve hand fatigue, carrying a price tag of $20,000 apiece.[3]

The Shuttle EMU (Figure 7.2) is NASA's most complex yet, with an

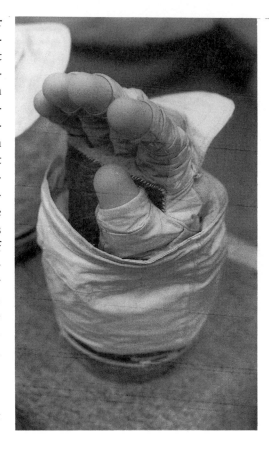

7.1
Gloves are the only part of the space suits still custom-sized to individual astronauts. (SI photo 85-6976-28)

equally sophisticated life support subsystem. Rather than an awkward, long body zipper, the new two-piece suit has metal rings that mechanically lock and provide pressure retention through the use of lip seals at the waist, helmet, and glove disconnects. Astronauts can efficiently don the Shuttle suit in five minutes by themselves. It took Apollo crew members as long as an hour to suit up, and they required assistance.[4]

Previous space suits became stiff when pressurized, and performing simple tasks or even grasping tools fatigued the wearer. To improve mobility, the molded convolute joints of Apollo suits were replaced by flat-patterned mobility joint construction in Shuttle suits. These low-fatigue

---

*What most people consider a "space suit" (the white body suit and gold visor, together with the backpack, or primary life support system) is designated the EVA suit or extravehicular mobility unit by professionals concerned with these items. For them, the term space suit refers only to those parts which cover the torso, limbs, and head, which excludes the primary life support system.

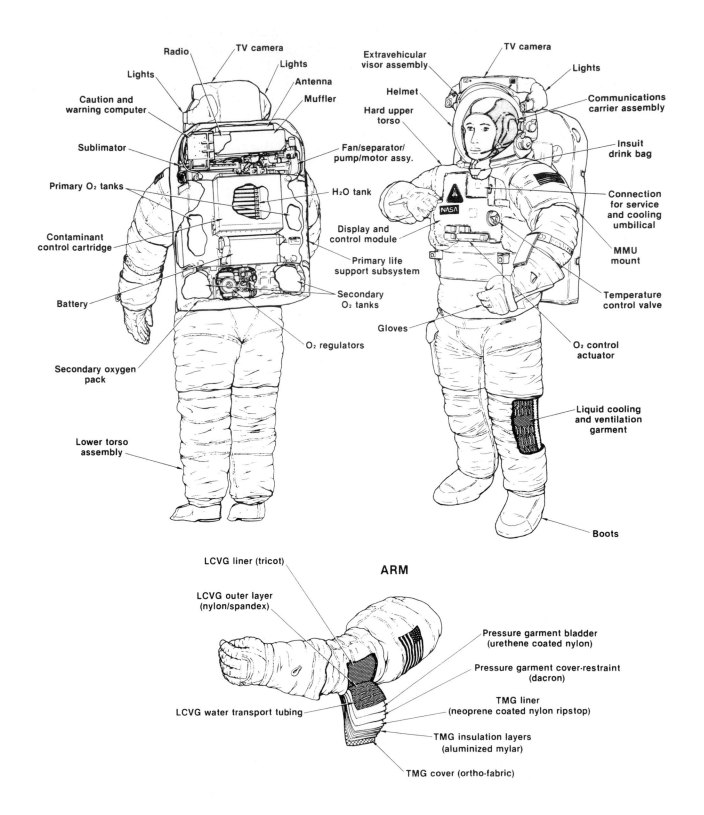

Radio

TV camera

Lights

Antenna

Lights

Muffler

Caution and warning computer

Sublimator

Fan/separator/ pump/motor assy.

Primary O_2 tanks

H_2O tank

Contaminant control cartridge

Display and control module

Battery

Primary life support subsystem

Secondary O_2 tanks

Secondary oxygen pack

O_2 regulators

Lower torso assembly

Extravehicular visor assembly

TV camera

Helmet

Lights

Hard upper torso

Communications carrier assembly

Insuit drink bag

Connection for service and cooling umbilical

MMU mount

Temperature control valve

Gloves

O_2 control actuator

Liquid cooling and ventilation garment

Boots

LCVG liner (tricot)

ARM

LCVG outer layer (nylon/spandex)

Pressure garment bladder (urethene coated nylon)

Pressure garment cover-restraint (dacron)

TMG liner (neoprene coated nylon ripstop)

LCVG water transport tubing

TMG insulation layers (aluminized mylar)

TMG cover (ortho-fabric)

mobility joints are designed for constant volume and low torque.*

The Shuttle EMU is worn only for work outside the Shuttle orbiter. The atmosphere inside the orbiter is close to airliner comfort, a shirt-sleeve environment of oxygen and nitrogen with a cabin pressure of 101.4 kPa (14.7 psi). The Shuttle's atmosphere can attain a maximum of 30-percent oxygen.

Inflight clothing for Shuttle astronauts consists of sky-blue trousers and jacket, both made of flame-resistant cotton isolated in Nomex cloth. A navy cotton-knit short-sleeve shirt coordinates with the inflight suit. The garments have built-in safety features and are fitted to be comfortable but not sloppy. Loose clothing could catch and inadvertently turn on a critical switch. Many closable pockets are available for storing small items that might otherwise float about dangerously in weightlessness. Stocked in specific pockets before flight are felt-tip and pressurized ball-point pens, mechanical pencils, data books, sunglasses, a Swiss Army pocketknife, and standard surgical scissors.

T-shirts and a pair of shorts are generally approved for sleep attire. Black masks and earplugs are available to help astronauts sleep. Shoes have little value in spaceflight because astronauts don't walk on any surface. Instead, soft wool slipper-socks are worn during orbital flight.

A special feature of the inflight suit is the waistband of the pants: Tabs are pro-

---

*Mobility joints maintain constant volume through the range of limb motions and provide restraint against elongation under pressure. Torque is a turning or twisting force. Flat-patterned mobility joints consist of folded pleats or tucks in the outer layer of the suit which allow easy flexing of joints without twisting, bunching, or stretching TMG material.

vided that allow the wearer to adjust the size as needed during a mission. At 1 gravity, valves in the blood vessels keep blood circulating correctly, but in space, of course, there is no gravity, and body fluids shift. Astronauts develop puffy faces because fluid shifts from the legs up through the thoracic cavity, arms, and head. Their clothes may no longer fit. Later upon reentry into earth's atmosphere capillary action restores fluids.[5]

Astronauts can leave the spacecraft, once the Shuttle is in orbit, to perform tasks outside the vehicle. Standard operating procedures call for two astronauts to perform most EVA operations. Before leaving the orbiter, each astronaut attaches a safety tether to his or her suit. This tether is a thin life line which the astronaut pulls on to maneuver alongside the spacecraft. A gentle tug is enough to change direction. The Gemini-style umbilical life line, of course, is not required.

EVA tasks beyond the near vicinity of the spacecraft, such as satellite retrieval, are facilitated by the manned maneuvering unit (MMU) (Figure 7.3), a Buck Rogers-style rocket chair. The MMU is operated by means of 24 thrusters that ensure precise control. The astronaut can hold a station-keeping position or maneuver up, down, sideways, in reverse, or forward by operating hand controls on the end of each armrest. The 117-kg (260-lb.) MMU is stowed in the front of the cargo bay. Some repair activities can be done in the cargo bay, where Shuttle EVA crews rely on a manipulator foot restraint. This restraint (Figure 7.4), which resembles a cherry picker and serves as a tool carrier and work platform, is part of the remote manipulator system (RMS).

To prepare for EVA, a crew member enters the orbiter's air lock where the

7.2 (Facing page) The EMU is capable of providing life support for Shuttle astronauts working in the space environment for up to eight hours. (illustration courtesy of the Hamilton Standard Division of United Technologies Corporation, SI photo 87-6781)

7.3
Bruce McCandless demonstrates the NASA manned maneuvering unit (MMU), which enables him to maintain a station-keeping position or maneuver up, down, sideways, forward, or in reverse. (NASA photo S82-27596)

7.4
Illustrating the "cherry picker" concept, McCandless uses the remote manipulator system (RMS) arm and the manipulator foot restraint (MFR). (NASA photo 84-HC-97)

SPACE SUIT/LIFE SUPPORT SYSTEM OR EXTRAVEHICULAR MOBILITY UNIT

1. LIQUID COOLING AND VENTILATION GARMENT Worn under the pressure and gas garment. Consists of liquid cooling tubes that maintain desired body temperature.

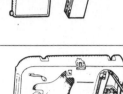

2. SERVICE AND COOLING UMBILICAL Contains communications lines, power, water and oxygen recharge lines and a water drain line. It has a multiple connector at one end which attaches to the EMU.

3. EMU ELECTRICAL HARNESS Provides bio-instrumentation and communications connections to the portable life support system

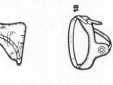

4. DISPLAY AND CONTROL MODULE Chest mounted control module which contains all external fluid and electrical interfaces, controls and displays.

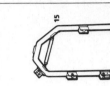

5. EXTRAVEHICULAR VISOR ASSEMBLY Attaches externally to the helmet. Contains visors which are manually adjusted to shield the astronaut's eyes.

6. HELMET Consists of a clear, polycarbonate bubble, neck disconnect and ventilation pad

7. ARM ASSEMBLY Contains the shoulder joint and upper arm bearings that permit shoulder mobility as well as the elbow joint and wrist bearing

8. GLOVES Contain the wrist disconnect, wrist joint and insulation padding for palms and fingers.

9. LOWER TORSO Consists of the pants, boots and the hip, knee and ankle joints.

10. HARD UPPER TORSO Provides the structural mounting interface for most of the EMU-helmet, arms, lower torso, primary life support subsystem, display and control module, and electrical harness.

11. PRIMARY LIFE SUPPORT SUBSYSTEM Commonly referred to as the "backpack," this assembly contains the life support subsystem expendables and machinery.

12. SECONDARY OXYGEN PACK Mounted to the base of the primary life support subsystem. It contains a 30-minute emergency oxygen supply and a valve and regulator assembly.

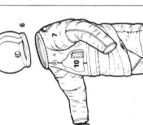
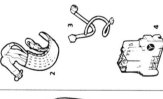

13. CONTAMINANT CONTROL CARTRIDGE Consists of lithium hydroxide, charcoal and filters which remove carbon dioxide from the air that the astronaut breathes. It can be replaced in flight.

14. BATTERY Provides all electrical power used by the space suit/life support system. It is filled with electrolyte and charged prior to flight. The battery is rechargeable.

15. AIRLOCK ADAPTER PLATE An EMU storage fixture which is also used as a donning and doffing station.

16. COMMUNICATIONS CARRIER ASSEMBLY Consists of microphone and headset. Allows the astronaut to talk to the other crewmen in the orbiter or other space suit/life support systems.

17. INSUIT DRINK BAG Stores liquid in the hard upper torso and has a tube projecting up into the helmet to permit the astronaut to drink while suited

18. URINE COLLECTION DEVICE Consists of the adapter tubing, storage bag and disconnect hardware for emptying liquid.

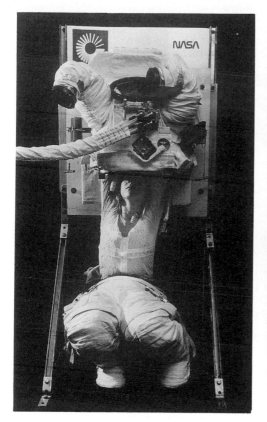 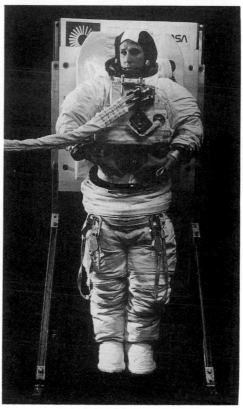

7.5 *(Facing page)*
Space Shuttle EMU and accessories. (illustration courtesy of the Hamilton Standard Division of United Technologies Corporation, SI photo 87-7727)

7.6
The lower torso is connected to the fiberglass upper section with a ring-shaped bearing. The service and cooling umbilical provides cooling water, power, oxygen, and communications for the astronaut prior to leaving the Shuttle. The modular suit and permanently attached backpack and display and control module greatly ease the donning procedure. (photo courtesy of the Hamilton Standard Division of United Technologies Corporation)

space suits are stored. The air lock is 1.6 by 2.1 m (5 ft. by 6.9 ft.) with a 1.0-m (3.3-ft.) diameter. D-shaped openings with pressure-sealing hatches allow suited crew members to transfer from the orbiter to space without depressurizing the entire crew compartment.

Once inside the air lock, the astronaut first puts on a urine collection device (UCD), followed by a liquid cooling and ventilation garment (LCVG)* and a com-munication carrier assembly (Figure 7.5). The astronaut then pulls on the trousers, or the lower torso assembly (LTA). The LTA consists of an integrated body seal closure, waist bearing, legs, and integral boots. The hard upper torso assembly, a vestlike rigid fiberglass shell, is mounted on the wall of the air lock. Crouching under the HUT, the astronaut reaches up into the arms to pull on the flexible sleeves, adjusting head and body into position (Figure 7.6). Secured to the inside front of the HUT is a drink bag with a 621-ml (21-oz.) capacity. The astronaut can sip water through a mouthpiece located at the top of the bag.

The LTA and HUT are connected by a pressure-sealed waist ring. A water line

---

*On early Apollo suits, oxygen vent-tubes had been integrated into the basic pressure suit. Late-model Apollo suits and Shuttle suits incorporated the oxygen-vent tubes into the liquid cooling garment. The suit liner was now technically called the liquid cooling and ventilation garment.

7.7 *(Top)*
Primary life support system components (worn on the astronaut's back). (illustration courtesy of the Hamilton Standard Division of United Technologies Corporation, SI photo 87-15036)

7.8
Display and control module components (worn on the astronaut's chest). (illustration courtesy of the Hamilton Standard Division of United Technologies Corporation, SI photo 87-15045)

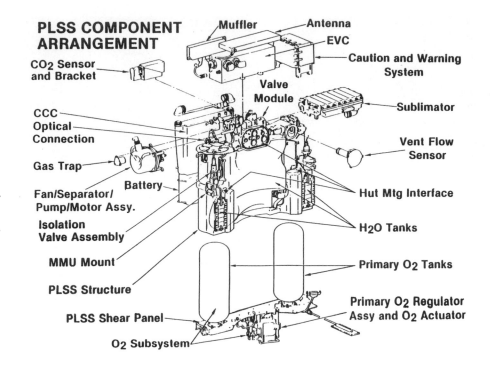

PLSS COMPONENT ARRANGEMENT

Muffler

Antenna

EVC

Caution and Warning System

Sublimator

Vent Flow Sensor

Hut Mtg Interface

H_2O Tanks

Primary O_2 Tanks

Primary O_2 Regulator Assy and O_2 Actuator

CO_2 Sensor and Bracket

Valve Module

CCC Optical Connection

Gas Trap

Battery

Fan/Separator/ Pump/Motor Assy.

Isolation Valve Assembly

MMU Mount

PLSS Structure

PLSS Shear Panel

O_2 Subsystem

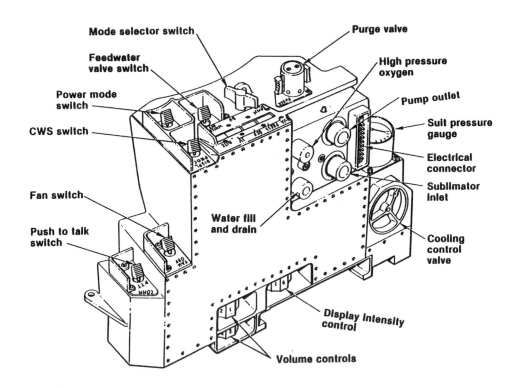

Mode selector switch

Purge valve

Feedwater valve switch

High pressure oxygen

Power mode switch

Pump outlet

Suit pressure gauge

Electrical connector

Sublimator inlet

Cooling control valve

CWS switch

Fan switch

Push to talk switch

Water fill and drain

Display intensity control

Volume controls

and vent tube inside the shell interior interface with the LCVG and the life support system (LSS). An umbilical to the chest area provides cooling water, power, oxygen, and communications prior to activating the primary life support system (PLSS) mounted on the rear of the HUT. The PLSS includes oxygen bottles, water tanks, fan/water separator/pump motor assembly, sublimator, contaminant-control cartridge, oxygen and water regulators, valves, sensors, and communications equipment (Figure 7.7). On Apollo suits, the PLSS was also carried on the astronaut's back, but was detachable. A secondary oxygen pack, (SOP) attaches to the bottom of the PLSS. The SOP provides 30 more minutes of pressure and breathing oxygen in case the suit develops a significant leak or the PLSS malfunctions.

The Shuttle EMU is also computerized. Apollo astronauts did have an electronic warning signal on their suits, but they had to make all adjustments manually. A microprocessor-activated caution and warning system on the Shuttle EMU's display and control module (DCM) (Figure 7.8) constantly tests vital functions of the EMU and instantly alerts the astronaut if something is amiss or expendables are low.[6]

When EVA astronauts go from a Shuttle cabin pressure of 101.4 kPa (14.7 psi) to a suit pressure of 28 kPa (4.2 psi), they risk the painful and potentially dangerous condition known as dysbarism (also called the bends or caisson disease). This condition arises when nitrogen gas, which is normally dissolved in the body tissues, comes out of solution in poorly vascularized tissues such as fat, connective tissue, and scar tissue. These nitrogen bubbles tend to collect near limb joints and produce pain. Danger occurs if large quantities of bubbles break loose and then travel to lodge in vital organs or the brain, causing loss of consciousness and death.

Twenty-four hours before the first EVA of a Shuttle mission, the two astronauts who will perform the EVA prebreathe pure oxygen for one hour to begin the process of denitrogenating the "slow" tissues of their bodies. The crew then reduces the orbiter cabin from 101.4 kPa (14.7 psi) to 70 kPa (10.2 psi) and 30-percent oxygen.

The space suit assembly (SSA) alone weighs approximately 47 kg (104 lb.). The primary life support system weighs 67 kg (148 lb.) and the helmet, lights, and camera weigh another 3.6 kg (8 lb.), bringing the total weight of the EMU to 117.6 kg (260 lb.). This would be a hefty weight to carry on earth, but in zero gravity, both astronaut and space suit assembly are weightless. The astronauts don their suits and float within the life-sustaining space suits breathing pure oxygen during the final forty minutes of EMU checkout. Pressure within the suit is initially 101.4 kPa (14.7 psi), but drops to 36 kPa (5.3 psi) when air in the Shuttle air lock is vented into space. When the air lock is at vacuum, the crew goes outside. The suit pressure drops from 36 kPa slowly because of leakage and metabolic conversion of oxygen to carbon dioxide by the crew members. When the suit pressure falls to 29.6 kPa (4.3 psi), the EMU's oxygen pressure regulators take over and keep suit pressure at that level for the remainder of the EVA. Sufficient oxygen is provided during EVA for breathing, and adequate body pressure is maintained, while the suit is kept from over-inflating and becoming stiff, which would hamper movement.[7]

Table 7.1
Shuttle Extravehicular Mobility Unit (EMU) Materials

Layer	Material	Function
	Thermal Micrometeoroid Garment (TMG)	
1	Ortho-fabric: Gore-Tex (expanded Teflon) fibers woven together with Nomex and backed with ripstop network of Kevlar	Abrasion/flame resistance
2–6	Aluminized Mylar backed with unwoven Dacron	Insulation
7	Neoprene-coated nylon ripstop	Liner
	Pressure Garment Assembly (PGA)	
8	Dacron woven with primary and secondary axial lines	Restraint and control of longitudinal growth
9	Polyurethane-coated nylon	Bladder layer for pressurization
	Liquid Cooling and Ventilation Garment (LCVG)	
10	Nylon acetate spandex woven with ethylene vinyl tubing	Restraint to keep tubing near body
11	Nylon acetate	Comfort

The first nine layers are integrated with the protective, vest-like fiberglass shell to form the hard upper torso assembly (HUT). For the lower torso assembly (LTA), the materials are integrated with the body seal closure, waist bearing, legs, and boots.

The PLSS provides approximately an eight-hour supply of oxygen for breathing, suit pressurization, and ventilation, and a 30-minute emergency life support supply. It begins functioning once the Shuttle umbilical is removed from the suit. After oxygen flows through the suit, it passes through a contaminant-control cartridge in the PLSS, containing a bed of lithium hydroxide that absorbs the carbon dioxide and converts it to lithium carbonate, plus water vapor and heat. Pure oxygen goes to the water separator, humidity is removed, and the oxygen passes to the fan which circulates it through the suit and PLSS. The oxygen next passes to the sublimator where it is cooled to 10°C (50°F) by sublimation of stored feed wa-

ter. The process is then repeated for the duration of the oxygen supply.

The pressure suit becomes fully functional in the vacuum of space. Woven between the nylon acetate spandex material and innermost nylon comfort layer of the liquid cooling garment, 91 m (100 yd.) of ethylene vinyl acetate tubing circulate cooling water to maintain the wearer's proper body temperature (Figure 7.9). Pressurized oxygen within the polyurethane-coated nylon bladder simulates the atmospheric pressure usually exerted on the body. The next layer is a woven Dacron restraint that limits radial growth of the pressurized bladder. The restraint layer also includes primary and secondary axial lines to control longitudinal growth while pressurized. A nylon ripstop material coated with neoprene rubber forms the lining layer of the thermal micrometeoroid garment. Five layers of aluminized Mylar, backed with unwoven Dacron, insulate the suit from shade-to-sun temperature extremes of −129° to 148°C (−200° to 298°F).[8]

Teflon was used in one form or another in different Apollo thermal micrometeoroid garments. The Skylab TMG used Teflon fabric in its outer layer. Gore-Tex expanded Teflon fiber was used for the new outer layer developed for the TMG between the Skylab and Shuttle programs. This Teflon fiber, called ortho-fabric, is a woven blend of Nomex and Gore-Tex fibers with a ripstop network of Kevlar threads. Gore-Tex expanded Teflon, or polytetrafluoroethylene (PTFE), was found to be five times more abrasion-resistant than plain Teflon. The slit-form construction of Gore-Tex yarn means that coverage is achieved using 226 g (8 oz.) of Gore-Tex fibers per square yard of fabric; conventional Teflon requires approximately 453 g (16 oz.). Gore-Tex fiber's natural white color reflects substantially more heat than conventional PTFE fibers, which are brown. Gore-Tex is also extremely temperature resistant in hot/cold chamber tests. Total fabric weight of the cover layer, made up of ortho-fabric, Nomex, and Kevlar, is 330 g/m² (14 oz./yd.²). On Apollo suits, joints were made of neoprene convolutions reinforced with cables; but Shuttle suits have flat-patterned mobility joints (similar to a tuck) stitched into the shoulder, elbow, wrist, knee, and ankle areas. This allows the joints to retain a flexed shape without constant muscle exertion.[9]

The Shuttle EVA helmet assembly consists of a transparent, impact-resistant polycarbonate shell, neckring, vent pad, purge valve, Velcro patch (to relieve itchy noses), and Valsalva device (to clear blocked ears). The vent pad assembly bonded to the inside rear of the shell diffuses incoming gas over the astronaut's face. The extravehicular visor assembly (EVVA), a light-and-heat-attenuating shell, fits over the helmet assembly. This assembly provides protection against micrometeoroid and accidental impact damage as well as solar radiation. A special gold coating on the sun visor reflects solar heat and light, yet permits the astronaut to see. Adjustable eyeshades provide further protection against sunlight and glare.

The EVVA also incorporates an EMU television camera and lighting attachment. Two lights provide sufficient illumination to use the miniature camera mounted in the housing to transmit pictures to the orbiter cabin and then to earth. Crew members in the cabin as well as ground personnel all share the astronaut's EVA view.

7.9
Iona Allen works on the Shuttle liquid cooling and ventilation garment (LCVG). An employee of ILC, Allen made Neil Armstrong's *Apollo 11* lunar boots—the boots that made historic footprints. (Photo courtesy of ILC Dover, Inc.)

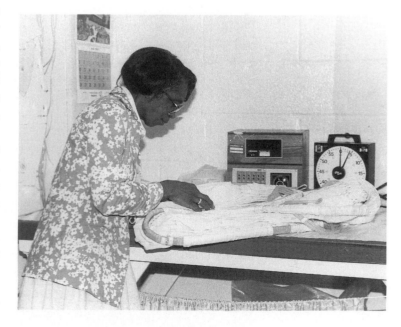

EVA gloves are made up of restraint and bladder layers encased in a TMG. The gloves are long enough to protect the astronaut's wrists as well as hands and are attached to the space suit arms at the wrist disconnects. A rotary bearing in the glove's wrist connector allows rotation and a wrist joint provides flexion/extension, as do fabric joints for thumbs and fingers. An insulating hot pad protects hands from temperature extremes.[10]

The Space Shuttle has a functional commode for both female* and male astronauts. But the waste management system for use under space suits is more complicated. For males, the system has not changed much since Mercury days. A urine collection device is attached to undergarments with Velcro and connected to the astronaut by a rubber sleeve (Figure 7.10). For EVA, female astronauts wear a garment similar to conventional feminine underwear called a disposable absorbent containment trunk (DACT). The DACT is lined with a transmission layer that conducts liquid waste to a very absorbent material that can hold more than a quart of liquid. Each crew member wears a DACT or UCD during ascent and reentry.

A pair of thick, tight-fitting shorts, known as a fecal containment system (FCS), had been in use aboard Apollo to contain solid waste and could be worn for longer EVAs. This system worked much like a disposable diaper. Once EVA was over, the astronaut removed the waste containment system and transferred the materials into appropriate containers. Experience indicated that this system was not widely used. The FCS was dropped from the Shuttle EMU quite early in the program.[11]

AFTERMATH OF *CHALLENGER*

The American Space program suffered a tragic setback with the loss of the orbiter *Challenger* and its crew of seven on January 28, 1986. Among the postaccident recommendations was a call to reexamine the feasibility of including emergency rescue and escape equipment on every Shuttle flight.

The first four Space Shuttle missions were test flights, and the crews were provided with special emergency equipment that was not standard on later flights. Rocket-propelled seats similar to those used in Lockheed SR-71 Blackbird reconnaissance aircraft provided the astronauts with a means for ejecting from a malfunctioning orbiter during launch.† The astronauts also wore U.S. Air Force high-altitude escape suits that were modified in respect to the way the parachute harness was attached (Figure 7.11).

From the fifth to the twenty-fifth Shuttle flights, there had been no means for crew escape during flight. Ejection seats and pressure suits were removed following the initial Shuttle trial flights because

---

*A contingency female urine absorption system (CFUAS) has been developed and successfully tested aboard the Shuttle as a backup in case of system failure. The Ellis Fitting, originally designed as an alternative to catheterization for female hospital patients, resembles a scoop with a tube leading off of it that connects to a highly absorbent collection bag.

†Once the Shuttle cleared the tower during these first four flights, astronauts could safely be ejected, in an emergency, up to an altitude of 30 km (100,000 ft.).

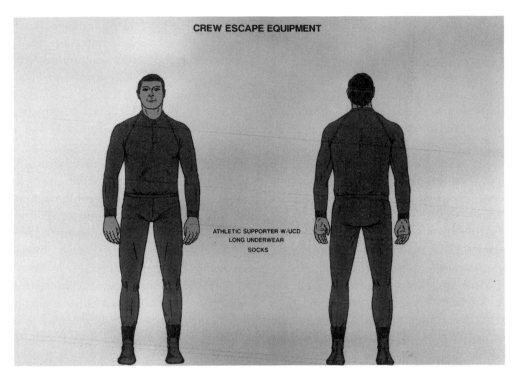

CREW ESCAPE EQUIPMENT

ATHLETIC SUPPORTER W/UCD
LONG UNDERWEAR
SOCKS

7.10 (*Top*)
Athletic supporter with urine collection device and long underwear with integrated socks. (SI photo 90-9102)

7.11
Shuttle crew escape equipment. (SI photo 90-9095)

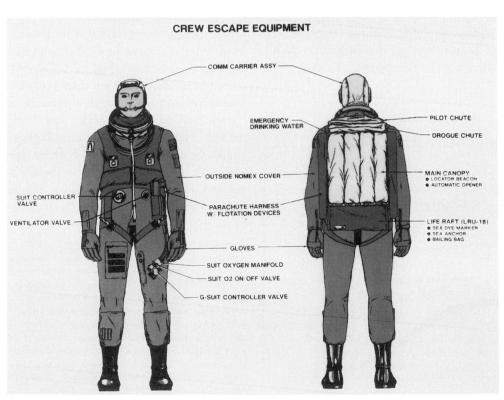

CREW ESCAPE EQUIPMENT

COMM CARRIER ASSY

EMERGENCY DRINKING WATER

PILOT CHUTE

DROGUE CHUTE

SUIT CONTROLLER VALVE

VENTILATOR VALVE

OUTSIDE NOMEX COVER

PARACHUTE HARNESS W/ FLOTATION DEVICES

MAIN CANOPY
• LOCATOR BEACON
• AUTOMATIC OPENER

LIFE RAFT (LRU-18)
• SEA DYE MARKER
• SEA ANCHOR
• BAILING BAG

GLOVES

SUIT OXYGEN MANIFOLD

SUIT O2 ON/OFF VALVE

G-SUIT CONTROLLER VALVE

these precautions were considered no longer necessary. However, many changes in safety practices were made following the loss of *Challenger*. If an emergency occurs aboard today's Shuttle, an escape hatch door jettisons. A long telescoping pole thrusts through the open hatchway and astronauts can hook onto the pole and quickly slide to safety beyond the Shuttle at which point their parachutes open (Figure 7.12). If the emergency is pre- or postlanding, an egress slide is deployed from the Shuttle side hatch that enables the crew to slide to the ground and run a safe distance from the Shuttle.

Most importantly, astronauts have returned to wearing the special one-piece partial pressure launch/entry suits (LES) similar to those worn aboard the first four Shuttle flights. The cover layer is bright orange for easy visual detection in a water emergency landing. The LES replaces the sky-blue, one-piece inflight coveralls used for launch and reentry in Shuttle flights five through twenty-five. Each LES is

equipped with oxygen, a parachute, a life raft, and other survival equipment. Additionally, astronauts wear communications gear, helmets,* and high-top boots (Figure 7.13).

Made by the David Clark Company, each escape suit has a comfort layer and a Gore-Tex inner liner integrated to a double-walled urethane-coated nylon gas container/exposure garment which either contains pressurized gas or can be manually inflated to create an air space that provides a thermal barrier against sudden exposure to air or water. Anti-g protection in the suits prevents blood from pooling in the lower extremities during reentry and the return to gravity, thus avoiding light-headedness or even temporary unconsciousness caused by hypoxia. Crew members inflate the legs of their g-suits just before deceleration. Bladders inside the suits squeeze the wearer's thighs and calves tightly to stop blood from flowing away from the upper body and head and settling in the legs. (Figure 7.14). Attached with snaps is a vent duct assembly that carries oxygen through a network of channels to the neck, hands, and down the legs (Figure 7.15). A restraint layer of stretch Gore-Tex linknet and Nomex tape keeps the gas container from overextending and aids pressurized mobility (Figure 7.16). Crew underwear consists of a cot-

7.12

Shuttle escape route. (SI photo 90-9105)

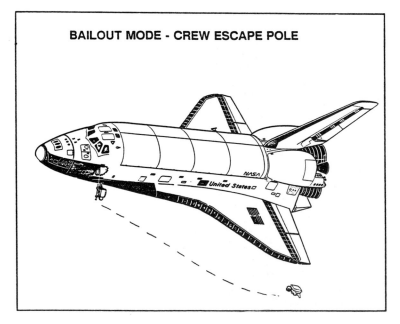

BAILOUT MODE - CREW ESCAPE POLE

---

*The launch ascent/descent helmet is a full pressure type with a fiberglass shell fitted with a large movable Plexiglas visor and connecting ring incorporating a thrust-type rotating bearing. The helmet is supported entirely about the shoulders with an adjustable formed wire support system that allows complete freedom of head movement. Astronaut John Young's helmet had a special feature. Prescription glasses had been mounted inside his helmet for use during orbital operations Other crew members have since also had optically corrected glasses placed in their helmets. This helmet or suit is not part of the EVA space suit.

ton long-sleeve turtleneck and long under-wear with integrated socks or briefs and socks.

The launch/reentry suit fulfills safety requirements for survival up to 30 km (100,000 ft.) in an atmosphere of −12°C (10°F) for 30 minutes and 24-hour survival in 7°C (45°F) water. Each suit weighs approximately 10.5 kg (23.5 lb.) and provides oxygen and full-body protection at extremely low barometric pressures until an optimum altitude is reached for bailing out of the orbiter. The LES integrates a counter-pressure suit and an anti-exposure suit into one unit. In case of sudden cabin pressure failure, the counter-pressure portion of the LES pro-tects against explosive decompression. Counter pressure applied is equal to the breathing pressure provided by the helmet. To avoid heat build-up experienced during launch and reentry, the suit can be plugged into a personal suit-ventilation system mounted onto the crew seats.[12]

SPACE SHUTTLE MISSIONS

Previous space missions focused on the technological refinement of the vehicles and support systems used to take humans into space. With the Shuttle program, the goals of spaceflight subtly shifted from exploration to operation. Astronauts would

7.13

Shuttle launch/entry suit. (SI photo 90-9093)

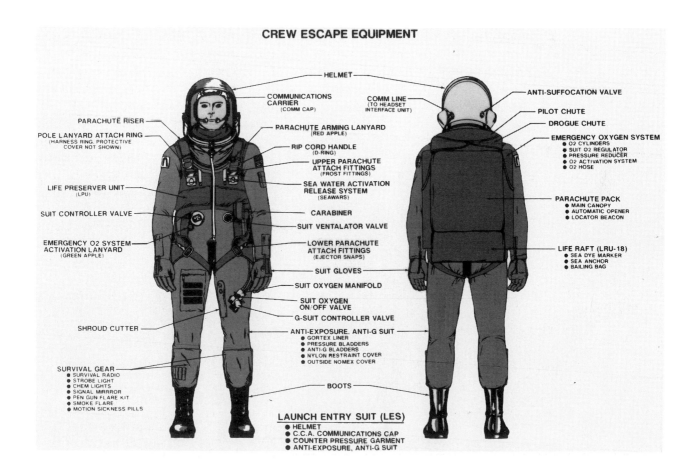

CREW ESCAPE EQUIPMENT

HELMET

COMMUNICATIONS CARRIER (COMM CAP)

COMM LINE (TO HEADSET INTERFACE UNIT)

ANTI-SUFFOCATION VALVE

PILOT CHUTE

DROGUE CHUTE

PARACHUTE RISER

PARACHUTE ARMING LANYARD (RED APPLE)

POLE LANYARD ATTACH RING (HARNESS RING, PROTECTIVE COVER NOT SHOWN)

RIP CORD HANDLE (D-RING)

EMERGENCY OXYGEN SYSTEM
● O2 CYLINDERS
● SUIT O2 REGULATOR
● PRESSURE REDUCER
● O2 ACTIVATION SYSTEM
● O2 HOSE

UPPER PARACHUTE ATTACH FITTINGS (FROST FITTINGS)

SEA WATER ACTIVATION RELEASE SYSTEM (SEAWARS)

LIFE PRESERVER UNIT (LPU)

PARACHUTE PACK
● MAIN CANOPY
● AUTOMATIC OPENER
● LOCATOR BEACON

SUIT CONTROLLER VALVE

CARABINER

SUIT VENTALATOR VALVE

EMERGENCY O2 SYSTEM ACTIVATION LANYARD (GREEN APPLE)

LOWER PARACHUTE ATTACH FITTINGS (EJECTOR SNAPS)

LIFE RAFT (LRU-18)
● SEA DYE MARKER
● SEA ANCHOR
● BAILING BAG

SUIT GLOVES

SUIT OXYGEN MANIFOLD

SUIT OXYGEN ON/OFF VALVE

G-SUIT CONTROLLER VALVE

SHROUD CUTTER

ANTI-EXPOSURE, ANTI-G SUIT
● GORTEX LINER
● PRESSURE BLADDERS
● ANTI-G BLADDERS
● NYLON RESTRAINT COVER
● OUTSIDE NOMEX COVER

SURVIVAL GEAR
● SURVIVAL RADIO
● STROBE LIGHT
● CHEM LIGHTS
● SIGNAL MIRRROR
● PEN GUN FLARE KIT
● SMOKE FLARE
● MOTION SICKNESS PILLS

BOOTS

LAUNCH ENTRY SUIT (LES)
● HELMET
● C.C.A. COMMUNICATIONS CAP
● COUNTER PRESSURE GARMENT
● ANTI-EXPOSURE, ANTI-G SUIT

7.14 *(Top)*
Pressure bladders, anti-g bladders, and Gore-Tex liner. (SI photo 90-9100)

7.15
Shuttle oxygen respiratory system. (SI photo 90-9097)

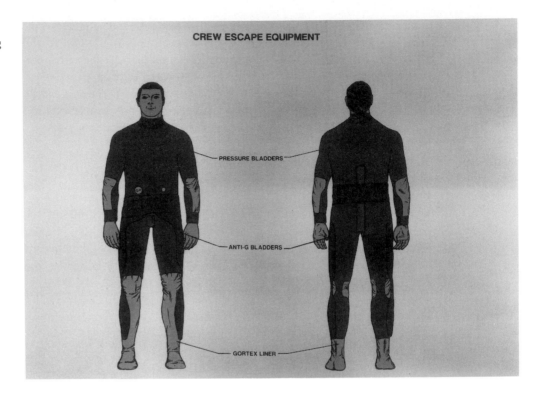

CREW ESCAPE EQUIPMENT

PRESSURE BLADDERS

ANTI-G BLADDERS

GORTEX LINER

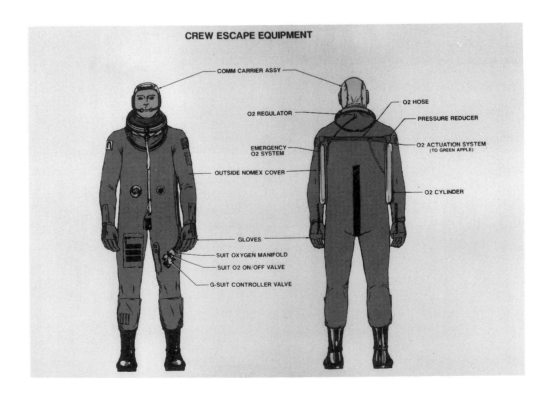

CREW ESCAPE EQUIPMENT

COMM CARRIER ASSY

O2 REGULATOR

O2 HOSE

PRESSURE REDUCER

EMERGENCY
O2 SYSTEM

O2 ACTUATION SYSTEM
(TO GREEN APPLE)

OUTSIDE NOMEX COVER

O2 CYLINDER

GLOVES

SUIT OXYGEN MANIFOLD

SUIT O2 ON/OFF VALVE

G-SUIT CONTROLLER VALVE

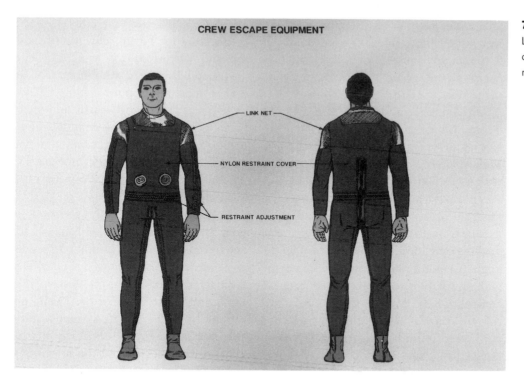

CREW ESCAPE EQUIPMENT

LINK NET

NYLON RESTRAINT COVER

RESTRAINT ADJUSTMENT

7.16
Linknet, nylon restraint cover, restraint adjustment. (SI photo 90-9099)

not simply be space explorers but rather trained workers in established facilities. Pilot astronauts were now joined by engineers, scientists, and technicians charged with determining the feasibility of working in space and investigating what new information could be found from that vantage.

Nine years after the last Skylab EVA, Story Musgrave and Don Peterson performed the first experimental Shuttle EVA in April 1983. During a three-and-one-half-hour EVA inside the payload bay area, they tested the new EVA space suits, tethers, tools, and EVA procedures. Both astronauts clearly enjoyed their successful venture, at times floating outside the payload bay, but were restrained from straying too far by their tethers. Musgrave,

comparing this experience to EVA training in the neutral buoyancy tank, thought this "tank" was a little deeper.

During Mission 41-B, in February 1984, Bruce McCandless and Bob Steward flew the new manned maneuvering units. For the first time, humans could orbit for brief periods without an umbilical or safety tether. McCandless, one of the MMU's chief designers, flew effortlessly 300 feet away from the orbiter and returned. McCandless compared flight on the MMU to that of flying a helicopter at Mach 25. Steward repeated the tests the next day. The two verified that the suits, primary life support systems, MMU, tools, and Canadian cherry-picker arm (remote manipulator arm) functioned correctly in preparation for repairs of the

Solar Max satellite on the next mission. Steward practiced refueling of the Landsat D spacecraft, planned for a later mission. He found the special tools easier to use in space than on earth.[13]

On Mission 41-C, in April 1984, George Nelson and James van Hoften began rescue operations for the ailing Solar Max. Nelson's initial attempt to dock with and stabilize the satellite was unsuccessful due to an unexpected grommet (in this context, a large bolt-like screw) projecting from the craft that prevented the docking operation. NASA Goddard Space Flight Center brainstormed to come up with a solution. They regained ground control of Solar Max so that the orbiter's robotic arm was able to grasp the satellite and bring it into the cargo bay where Nelson and van Hoften repaired Solar Max and successfully released it back into orbit. This opened the door for astronauts to undertake other rescue missions. To successfully grapple shiny, new satellites in space, worth millions of dollars but technically lost because their motors failed, repair them, and restore them to proper orbit could save insurance underwriters a great deal of money. It could also demonstrate the unique adaptability of humans and the MMU.

Kathryn Sullivan became the first American female to walk in space in October 1984, on Mission 41-G. Speaking of the Shuttle suit, she explained, "It does not have all the same ranges of motion that your body has. Its shoulder is not like yours, and its knee is not built like yours, so you learn, through several runs in the water tank, a whole set of lessons that have to do with suddenly being a person of greater mass and volume. For instance, you can't put your nose right up to something." During their EVA, Dave Leestma

and Sullivan practiced satellite refueling in preparation for future refueling of Landsat and earth-resource satellites. They also manually stored stubborn radar antennae in preparation for reentry. Sullivan remembers the view of earth and space, no longer limited by the windows of the orbiter's cabin, as spectacular and panoramic as the universe itself. But the EVA schedule is too tight to allow for much awed sightseeing.[14]

Because of the successful retrieval, repair, and relaunch of Solar Max, NASA convinced Lloyd's of London to underwrite retrieval of Palapa and Westar. Dale Gardner and Joe Allen, on Mission 51-A in November 1984, used the MMU and robotic arm to track down and capture two satellites. Anna Fisher, the first female medical doctor and the first American mother in space, guided the arm from within the Shuttle. Gardner and Allen used the two tools in their recovery attempt. Allen compared the ease of control an astronaut had while flying the MMU to that of a pilot in an enclosed plane. By contrast, he compared the nerve-racking ride on the robotic arm high above the cargo bay to standing on the world's highest diving board. The danger of a free fall to earth from a movable perch seemed very real to Allen even though he knew better. For recovery of Palapa B2 and Westar VI comsats (communication satellites), Lloyd's of London later presented the three astronauts the Lloyd's Silver Medal, making Anna Fisher only the second woman to receive this award.

An unplanned EVA took place in April of 1985, during the flight of 51-D, when the Syncom 4-3 comsat failed to activate. David Griggs and Jeffrey Hoffman went outside the orbiter to tape handmade devices to the robotic arm. Rhea Seddon, inside the Shuttle, jockeyed the robotic arm to make contact with the stubborn Syncom. She made several successful contacts with the craft's lever, but it still failed to function. Analysis of photos later determined that the failure was inside the Syncom and had nothing to do with the external lever. Engineers on ground control would have to study the Syncom problem until another repair attempt could be made.

NASA wanted to demonstrate Shuttle's flexibility, and the builder of Syncom, Hughes Aircraft Corporation, wished to recoup as much as it could from the malfunctioning satellite. Bill Fisher and Jim van Hoften, in August of the same year, stood ready in the payload bay of STS-51-I to begin capture and repair operations. It took seven hours for Fisher and van Hoften to accomplish their repairs with the aid of Mike Lounge, who operated the cherry-picker arm from within the Shuttle. The next day, van Hoften completed the assignment by projecting Syncom out into space once more. Ground control reported Syncom functioning successfully in orbit.

For Mission 61-B in late 1985, Jerry Ross and Sherwood Spring spent a total of 13 EVA hours during two days, practicing erecting and dismantling framework nicknamed EASE (Experimental Assembly of Structures). The two astronauts, floating freely, created an inverted pyramid framework. Spring complained about aching hands; astronauts are always concerned about fatigue from long space walks and work in space. After a day's rest, the two returned to their construction practice. With Mary Cleave's help inside the spacecraft, Ross used the robotic arm to assist him in topping off the structure. The astronauts' work increased understanding

of space engineering, providing designers enough information for years of space station studies.

The *Challenger* accident temporarily stopped Shuttle space travel. Two and one-half years were to elapse before intensive efforts by NASA and its contractor teams paid off. With flight safety as the primary goal, $2.4 billion was spent on redesigning and replacing critical components of the Shuttle fleet. More than 400 modifications were made to the Shuttle and its propulsion systems. Crew members have the additional protection of their orange launch/entry suits, now mandatory for those phases. The crew escape system allows for safe exit from a disabled orbiter on the pad, in flight, or on the ground. With durable, high-temperature-resistant launch/entry suits, the crew can survive conditions of high-altitude decompression as well as the environments encountered at sea level.

The launch of *Discovery,* STS-26, September 29, 1988, and its completed mission culminated efforts to develop effective, safe, and reusable space transportation. Other successful Shuttle flights have followed and other important satellites have been deployed, establishing that the Shuttle has the ability to provide significant space benefits to the nation.[15]

More than five years after the last EVA activity, STS-37 mission *Atlantis* lifted off April 5, 1991, carrying Steve Nagel, Ken Cameron, Jay Apt, Linda Godwin, and Jerry Ross. On an unscheduled four-and-one-half-hour EVA, Ross and Apt freed a balky antenna on the observatory satellite, Gamma Ray Observatory (GRO), enabling the $617-million device to be placed in orbit. GRO, the most massive science satellite yet launched, was also designed for propellant refueling. Scientists hope the multi-instrumented spacecraft will explain some of the universe's most explosive objects and prove theories of exotic phenomena, such as quasars and gamma ray bursts. While Godwin operated the remote manipulator system that lifted the satellite out of the payload bay, the accordion-fold solar array panels were slowly deployed. Later on, when Goddard Space Center's repeated commands failed to activate GRO's high-gain antenna into position, mission control sent Ross and Apt on EVA to perform deployment procedures.

For their second EVA, the astronauts tested sleds proposed for hauling people and large objects through space on a rail in anticipation of a working space station. While practicing with the sleds, Apt accidentally pierced the pressure bladder layer of one glove. The hole was not discovered until his gloves were inspected following the flight because, fortunately, the puncture did not leak substantially. The hole was caused by the stainless steel palm bar designed to prevent ballooning of the pressurized glove. The palm bar also pricked Apt's hand, drawing a small amount of blood. As for the sleds, Ross and Apt decided that less sophisticated systems worked best. For moving along a monorail, they preferred a simple manual cart and the Shuttle tether. Apt said the cart takes very little strength: Two pulls would almost take one to Mars. And that is exactly where the astronauts hoped to go!

Besides a manned exploration of Mars, plans for future space endeavors call for unmanned satellite studies of earth's environment and atmosphere and a permanent manned base on the moon. Scientists and researchers are studying proposals for space stations, as well as substantial and long-range programs to support such plans. The need for space suits for EVAs long enough to maintain space stations still intrigues suit engineers and designers.

● ●

The Future of Space Exploration

The cold war ended in 1990. Soviet troops came home from Afghanistan. The Berlin Wall tumbled down, and the divided Germanies became one. As the heavy rod that supported the iron curtain fell, Soviet satellite countries declared their independence one by one. The United States and the former Soviet Union turned to focus on the many problems of their homefronts.

Walter McDougall says in *The Heavens and The Earth,* that political patterns of space technology are in great flux, particularly as that technology continues to diffuse. New nations are staking claims on the high frontier as the roar that was the wave of the twentieth century ebbs away. What will the twenty-first century bring? Perhaps it will be as the Japanese National Space Development Agency (NASDA) President Masato Yamano says, no longer a case where just one or two countries promote their own space development. Rather exploration of space will become a joint venture of humankind.[16]

● ●

CHAPTER 8

Advanced Development Space Suits

Looking ahead to permanently manned space stations, NASA considered space suits that would incorporate a hard suit concept. Suits worn by Mercury, Gemini, and Apollo astronauts were soft and fabricated almost entirely of flexible materials. Shuttle suits are hybrids, a direct result of advanced suit development. Some parts of these suits are made of hard materials and some are made of softer fabrics, films, and elastomers (elastic polymers, including rubber). A hard suit can be put on quickly for emergencies, such as extravehicular repairs, and it will work at higher pressures, thus reducing the time spent prebreathing now required before astronauts can perform EVA. Space station astronauts will likely perform longer and more frequent periods of EVA and will require even more protection from their hostile working environment. NASA bioengineers believed that the long-duration assembly should be a hard suit, as it would be more durable, rugged, and reliable. The space agency was keenly interested in Litton Industries' work on such a product.[1]

Development of the hard suit was pioneered in 1955 by Siegfried Hansen at Litton. Hansen had been one of several electronics engineers who left Hughes Aircraft Research Laboratory when Howard Hughes decided to pursue government contracts rather than the development of advanced electronics. A group of the engineers purchased a small

8.1

America's first space travel-simulation laboratory, capable of evacuation to pressures encountered several hundred miles above the earth, was built under USAF sponsorship for Litton Industries, Beverly Hills, California. (SI photo 91-1950, from Siegfried Hansen's scrapbook, courtesy of Litton Industries, Inc.)

microwave-tube company owned by Charles Litton. This company became Litton Industries Space Research Laboratory.

Hansen was developing an inhabitable evacuated chamber in which to assemble, modify, and test vacuum tubes. The chamber, custom-fabricated by the Lacy Manufacturing Company, was a 4.6- by 2.4-m (15- by 8-ft.) steel plate rolled into a cylinder and lying on its side. This vacuum chamber simplified the study of electron tubes. Normally, an electron tube contained a collection of grids and filaments arranged in certain patterns and en-

closed in a glass envelope from which all air had been pumped. Each time after components were set up inside the glass envelope, the air had to be pumped out. Litton's chamber expanded the glass envelope to allow researchers inside the vacuum (Figure 8.1).

Protective clothing was needed for engineers working inside the vacuum. Available pressure suits lacked the mobility the job demanded. Hansen and Litton technician Allan De Vantine were in charge of designing a new suit. The resulting product was the Mark I suit (Figure 8.2), which was later refined with NASA's support and evolved into the RX series of hard suits.[2]

Before *Sputnik,* the popular press ridiculed the concept of man in space. But the Air Force, realizing the coming importance of space, had agreed to fund the Litton project. The contract was subject to one clause—the terms "space" or "space suits" must never be mentioned. The Air Force told Litton to refer to the lab as the inhabited High Vacuum Laboratory and progress reports referred to a "manipulator station." Cartoons of the hard suits appeared as early as 1958 (Figures 8.3 and 8.4), while real flight surgeons and Litton engineers tended to the original vacuum chamber.[3]

The suit, a combination of rubber and fabricated aluminum parts, weighed 22.7 kg (50 lb.). The inner suit, a modified double-walled Air Force ventilation suit, distributed oxygen over the entire body. The trousers were cannibalized from an old Navy Arrowhead model AX-6 pressure suit. To avoid the ballooning effect of full pressurization, the suit had jointed aluminum ribbing that gave the wearer mobility. Rubber vapor jackets over the hands served the same purpose as gimbal-jointed arms and umbilical. The helmet was rigidly attached. The outer suit, or vapor suit, was made of a special natural rubber compound free of all volatile elements that could corrupt the purity of the vacuum.

In laboratory language, the person wearing the suit was the "inhabitant." For his tests, Hansen chose to be the first inhabitant (Figure 8.5). Hansen would prebreathe oxygen for approximately one-half hour to rid his system of nitrogen to prevent the bends. Then he would don the suit over long red underwear. Oxygen would be pumped into the suit at 34.5

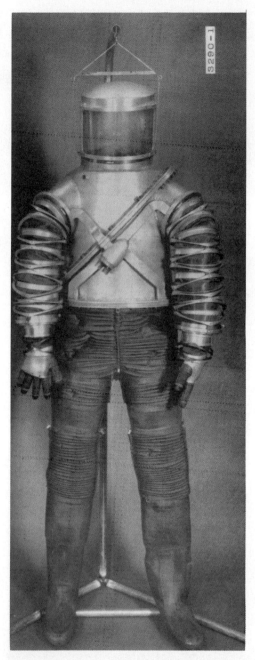

8.2

The Litton Mark I suit was developed in 1957 by Siegfried Hansen under a USAF program. It was the first American hard suit to achieve constant volume pressure and good mobility through the use of rigid restraining rings, or rolling convoluted joints. This hard suit program later evolved into the NASA-supported RX series of suits; and the Mark I and RX series of hard suits evolved into the Shuttle space suit. The future space station suit will probably be a spin-off from this hard suit technology. (SI photo 86-14174)

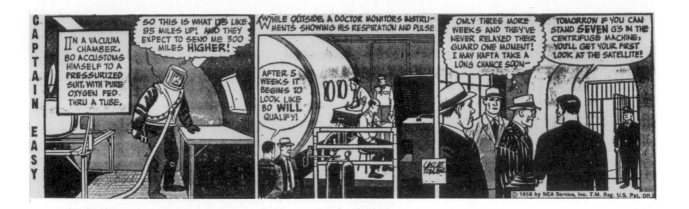

'Candidates? I Was Elected Years Ago'!

kPa (5 psi), and engineers would bring Hansen to a simulated altitude of about 8,229 m (27,000 ft.) (Figure 8.6). The suit was air-cooled with the same oxygen used for breathing. As the air inside the suit heated up, it would be drawn off and passed through refrigerated brine.

Hansen usually began experiments by dropping two feathers. They would fall like lead lumps. He also lit a thermionic tube (a vacuum tube in which a flow of electrons is emitted by heated electrodes) without its enclosing glass envelope, proving the absence of air. The cost of running an inhabited chamber test averaged $125 per hour.

Some of Litton's research had been concerned with development of the Chromatron color television tube. Other programs related to the new space program. Researchers worried that a spacecraft landing on the moon would sink into the powdery surface. Hansen's group demonstrated that a heavy steel ball dropped in earth's atmosphere did sink to the bottom of a container of powder, but in the high-vacuum chamber, the dropped ball only indented the surface of the powder. The safe Apollo landings on the moon proved their findings correct.

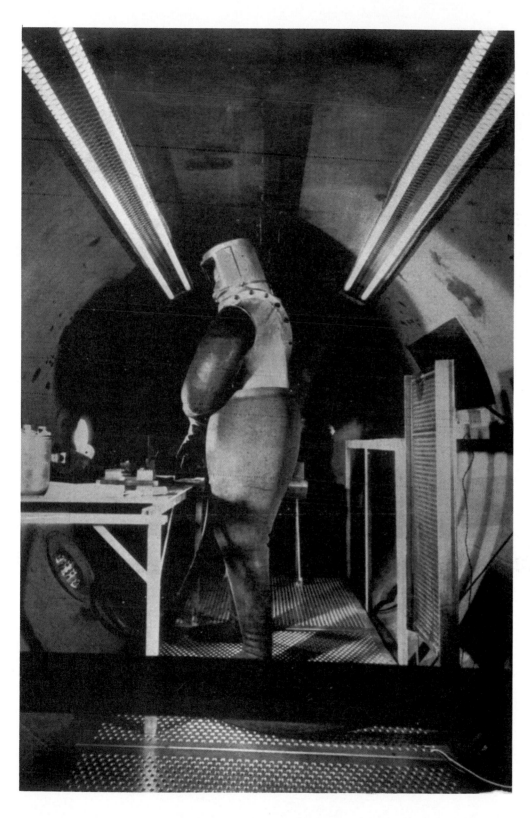

8.3 *(Facing page, top)* It appears that the artist for "Captain Easy" did his homework in 1958. The suit and chamber strongly resemble Dr. Hansen's suit and Litton's sealed chamber. ("Captain Easy" reprinted by permission of NEA, Inc., SI photo 91-1954, from Siegfried Hansen's scrapbook)

8.4 *(Facing page, bottom)* On April 9, 1959, the list of Mercury astronauts was announced. This cartoon showing Dr. Hansen's face inside a Litton suit with a vacuum tube-type helmet appeared in a Seattle newspaper. ("Captain Easy" reprinted by permission of NEA, Inc., SI photo 91-1957, from Siegfried Hansen's scrapbook)

8.5 *(Left)* Dr. Hansen, as the inhabitant, wears a 22.7-kg (50-lb.) aluminum and steel "space suit" in the sealed vacuum chamber. Hansen set a record by remaining at a simulated altitude of 152 km (95 mi.) for approximately three and one-half hours. In this "moon room" he demonstrated the use of screw drivers, pliers, and power drills and performed inert gas welding. (SI photo 91-1971, from Siegfried Hansen's scrapbook, courtesy of Litton Industries, Inc.)

RX SERIES

The RX series of suits, evolving from the Mark I, was manufactured by Litton Systems, Inc. These suits consisted of articulated, exoskeletal, anthropomorphic enclosures with segments united by joints to form an external pressure vessel. The ideal hard suit could maintain a nearly perfect constant volume throughout the full range of body motions. A hard suit could also operate at higher pressure, re-

ducing the time-consuming need to prebreathe oxygen before EVA.

The key to solving the constant volume problem was Hansen's rolling convolute. Earlier suits, like Colley's tomato-worm model, used convoluted joints (bellows) to provide mobility, but these suits often failed to maintain a constant volume over the joint's range of motion, with resulting high torques and "spring back." The rolling convoluted joint allowed the joint fabric to roll in a controlled and volume-

8.6

Litton flight surgeons and engineers carefully monitor Dr. Hansen in the vacuum chamber. (SI photo 91-1958, from Siegfried Hansen's scrapbook, courtesy of Litton Industries, Inc.)

compensated manner through the complete range of movement. Previous RX suits had used the rotary joints (bearings) only in the wrists. Hard suit joints, sometimes called stovepipe joints, consisted of angular cylinders of pipe with rotary joints on each end. Rotary joints consisted of an inner and outer aluminum bearing race (or groove) with ball bearings in between.

The suit was sealed not by zippers, but by a "doglock," something like the fixture used on a submarine hatch. It was thought that the hard suit provided not only a lower leakage rate, but also more abrasion resistance and better protection against micrometeoroids.[4]

The RX-1 suit (Figure 8.7) was delivered to NASA in early 1964. It provided excellent arm, shoulder, and leg mobility at 34 kPa (5 psi). Overall mobility was hampered by its weight of 37 kg (83 lb.)* and a restrictive soft-type hip section. However, this suit provided a test assembly for mechanical and structural features and for the evaluation of the constant-volume joints.

The RX-2 (Figure 8.8) was ready for the agency in December 1964. Hard pants with constant-volume hip joints were substituted to solve the hip articulation problem. The RX-2 showed a marked improvement, but the user was handicapped by a suit weight of 40 kg (90 lb.), lack of waist mobility, and a relatively high center of gravity that caused some instability.

Completed in June 1965, the RX-2A differed considerably from earlier models. Weight was reduced by 4.5 kg (10 lb.) and bulk was reduced in the shoulder breadth

*The then "state-of-the-art" Mercury suit weighed 9 kg (20 lb.).

from 73 to 58 cm (29 to 23 in.). The shell consisted of a composite aluminum and fiberglass honeycomb structure, with redesigned dual-plane body seals rather than the single-plane diagonal closure formerly used. The redesigned, fully articulated waist joint more closely matched the fore-and-aft and side-to-side bending motions of the human waist. A precise semi-circular groove was machined on the inside face of the lower flange of the body

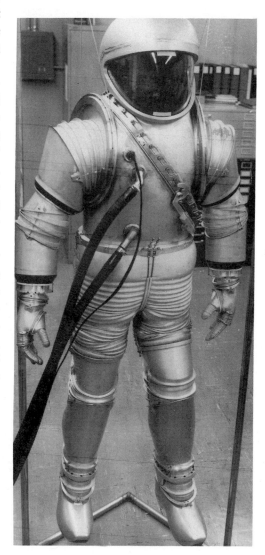

8.7
The RX-1 suit evolved from the early Mark I in 1964. Arms, legs, and shoulders reflected greater mobility. (SI photo 91-1951, from Siegfried Hansen's scrapbook, courtesy of Litton Industries, Inc.)

8.8 (*Left*)

The RX-2 came in December 1964. It solved the difficult hip articulation problem by substituting hard pants and constant-volume hip joints for the soft section between waist and knees. Movement was still hampered by a suit weight of 40 kg (90 lb.) and a lack of waist mobility. (SI photo 84-10720)

8.9

For certain elements of the RX-4, a standard size was made. Other elements, such as the legs and arms, could be replaced by larger or smaller parts. (SI photo 84-10719, from Siegfried Hansen's scrapbook, courtesy of Litton Industries, Inc.)

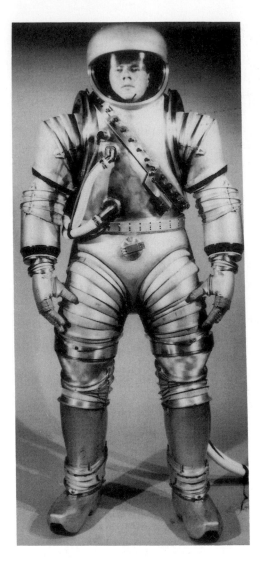

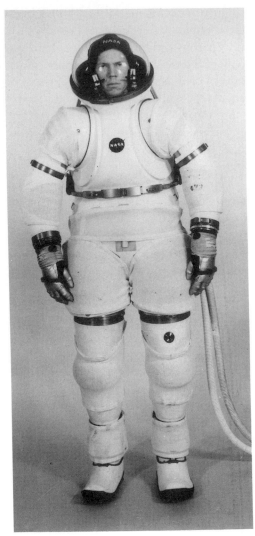

seal closure. The upper ring of the fully armored midsection shell (waist joint) carried matching grooves on its outer face. When these grooves were placed in opposition to the grooves provided on the closure flange, they formed a tunnel through which an equal length of flexible wire was drawn. The two telescoping parts then locked into any desired position.

Also included were a hemispherical helmet and a modified shoe contour. A unique characteristic of the RX series is the shoe, which was patterned after the Dutch sabot. Litton engineers purchased a pair for $4.90 and concluded that the sabot had the proper sole and heel contour for walking on the lunar surface. They used the Dutch shoe, with the toe shaved off, for a mold and added polyurethane insulation in the sole to form the hard boot.[5]

NASA's Manned Spacecraft Center in Houston received the RX-3 in 1966. It weighed 28.5 kg (62 lb.), a noteworthy

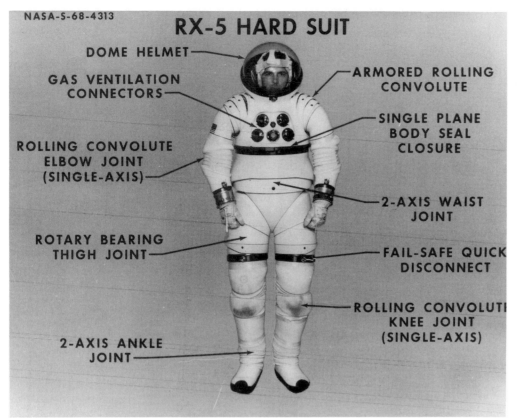

NASA-S-68-4313

RX-5 HARD SUIT

DOME HELMET

GAS VENTILATION CONNECTORS

ARMORED ROLLING CONVOLUTE

SINGLE PLANE BODY SEAL CLOSURE

ROLLING CONVOLUTE ELBOW JOINT (SINGLE-AXIS)

2-AXIS WAIST JOINT

ROTARY BEARING THIGH JOINT

FAIL-SAFE QUICK DISCONNECT

ROLLING CONVOLUTE KNEE JOINT (SINGLE-AXIS)

2-AXIS ANKLE JOINT

8.10
The RX-4 and RX-5 suits (1968) proved excellent for routine and experimental testing. Both shared similar features with slight development changes and reflected improved joint mobility and an improved helmet. (SI photo 84-10714)

7.5-kg (18-lb.) reduction accomplished by using magnesium and magnesium-lithium alloys in the rotary seals and shell elements. Design goals concerning torque, mobility, thermal protection, and defense against radiation and micrometeoroids were all met. The RX-3 also had a versatile sizing system. Limb and torso elements could be made available in a wide variety of standard sizes.

In place of the liquid cooling garment, engineers wanted to design a thermal protection system for the hard suit based on heat exchangers to maintain an air-conditioned effect. This would reduce donning time and eliminate the necessity for a backpack, by miniaturizing the PLSS and making it an integral part of the suit. The suit's center of gravity would be low-

ered, adding more stability for the wearer. The RX-3 suit was a feasibility demonstrator for the subsequent development of a flight-qualified suit.

The goal of the RX-4 suit (Figure 8.9) was to provide a prototype that could qualify for lunar exploration missions. Engineers now called for a space-compatible helmet-sunshade assembly, as well as compatibility with the Apollo portable life support system and liquid cooling garment. Litton also studied ways to nest the suit sections so that they would fit into a small storage area of the Apollo spacecraft.

Miscellaneous modifications were made to improve the RX-4 design. The RX-5 (Figure 8.10) further reduced the weight from 28 to 22 kg (63 to 50 lb.). An

8.11

The RX-5 soft suit (1969–70), Litton's advanced extravehicular suit (AES), consisted of two main sections broken at the middle by a quick-disconnect torso rotary seal *(far left)*, which along with convoluted joints, keeps suit volume constant, no matter what position the wearer assumes. At right are front and rear views of the suit without its cover layer. (SI photo 87-15048)

increased mobility range in the waist and the knee and hip joints enhanced kneeling capabilities and leg movements in general. Designers also provided fail-safe quick disconnects, and as many standardized equipment items were used as possible to provide compatibility with the Gemini and Apollo programs. The basic hard suit structure was a composite layup of a thin aluminum sheet faced with a fiberglass honeycomb section for stability.

An outgrowth of the RX-5 hard suit program was a second suit of similar design but constructed of non-flammable material. This RX-5 soft version suit, Litton's advanced extravehicular suit (AES), was fabricated as two main sections joined by a quick-disconnect system at the waist (Figure 8.11). The RX series incorporated armored rolling convolutes (pro-

tected by a metal covering) for most joints, but the AES (Figure 8.12) sported a multiple-bearing concept in the shoulder and hip areas (Figure 8.13), with armored rolling convolutes at the elbows (Figure 8.14), waist, knees, and ankles. A hemispherical helmet completed the suit.

RX-4 and RX-5 hard suits were particularly suitable for the chamber use which inspired Siegfried Hansen's original development work. Technicians could now perform experimental procedures, such as electron-beam welding of large assemblies, in chambers that simulated the conditions of space. The hard suit exoskeleton offered greater safety because of its invulnerability to accidental tearing or puncture. Litton engineers believed that only an exoskeleton could support the human frame efficiently during sustained ac-

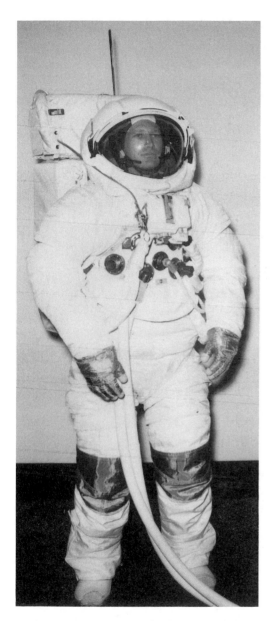

gram are now reflected both in the current Shuttle suit assembly and in advanced space suit designs. These include the hard upper torso and hard suit sizing elements, the single- and dual-plane body seal closures, bearing joint and gimbaled-joint systems, and modular element sizing.[6]

The cost of the RX hard suit program was $2.8 million. The total spent from 1962 through 1969 for the development of hard suit technology, from Litton's Mark I through a 55-kPa (8-psi) orbital extravehicular suit (OES), was approximately $8 million (or about 13 percent of the approximate $62-million cost of the Apollo suit program).[7]

8.12

The fireproof RX-5A soft suit (1969–70), an AES, had an integrated fireproof thermal micrometeoroid garment. This suit reflects an outgrowth of Litton's RX-5, mainly used to prove the mobility of the RX-5 joints in a fabric suit. The USAF considered the AES as a possible contender for use in the Manned Orbiting Laboratory. (SI photo 86-14177)

celeration such as that experienced during an ejection bailout. The hard suit was even similar, in essence, to a spacecraft in that systems such as environmental control were integral parts, thus reducing weight and bulk.

Many significant development objectives derived from the RX hard suit pro-

AX SERIES

The NASA Ames Research Center also produced hard space suits, designated the AX series (Figure 8.15), in conjunction with the AiResearch Division of the Garrett Corporation. AiResearch studied existing hard suit technology very thoroughly, even contacting the Tower of London to inquire about the finely articulated joints in medieval suits of armor (Figure 8.16, notice neck articulation). AiResearch decided upon multiple bearing technology for its 1966 model, the AX-1. Model AX-2 (Figure 8.17) included rotary closures and an increased waist flexion range. Seeking maximum mobility, long operational life, and low-cost fabrication, the AiResearch advanced extravehicular suit used new toroidal (doughnut-shaped) joint technology exclusively (Figure 8.18), except for a multiple-bearing shoulder joint.[8]

The AX-3 suit evolved from a review of other assemblies, component developments, and mobility exercises. It had a

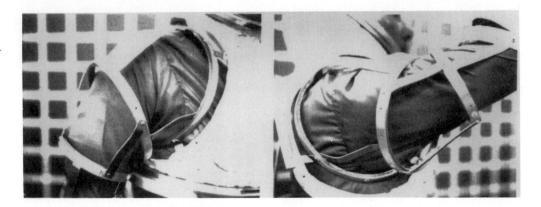

8.13 (*Top*)
Litton shoulder rotary seal development assembly for the fabric suit, June 1967. (SI photo 86-14834)

8.14
Litton prototype elbow joint for the fabric suit, June 1967. (SI photo 86-14836)

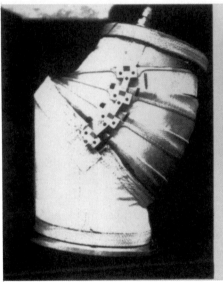

hard torso and all-soft components consisting of a multiple-laminate structure of neoprene-coated Nomex and ripstop materials. The shoulder joint had three sealed bearings with an internal linkage and a tapered rolling convolute arrangement. The torso incorporated hard structures above and below the dual-plane entry closure and in the section between the waist and hip joint (Figure 8.19). This configuration allowed the maximum area to be utilized on the rear of the suit for mounting the life support system and provided ease of donning. The joint at the hip and thigh had a sealed bearing combined with a soft rolling convolute. Interchangeable rings of varying lengths provided a large range of suit sizes (Figure 8.20). A universal-fit boot eliminated the need for custom boot liners. Instead, a leather strap laced around the ankle and instep kept the boot in place.

Hubert C. Vykukal, principal engineer in crew and human factors research at

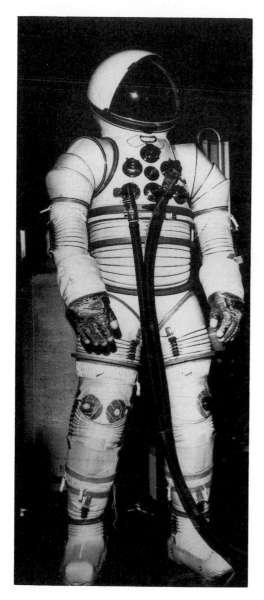

8.15 (*Left*)
The AiResearch lunar advanced Apollo program (LAAP) hard suit (late 1960s) was intended for *Apollo 18, 19, 20,* and *21,* prolonged lunar missions that were scrapped. (SI photo 84-10713)

8.16
In the 1950s and 1960s space suits began to resemble suits of armor. Serious research into the joint articulation of armor had implications on the development of the RX and AX series of hard suits. This armor was made for George Clifford, Third Earl of Cumberland, circa 1590. (photo courtesy of the Metropolitan Museum of Art, New York, Munsey Fund, 1932, 32.130.6 a-y; SI photo 91-1967)

Ames, worked on hard suit designs since the AX-1 (late 1970s). He designed the AX-5 (1980s), nicknamed the "Michelin Man" because of its eye-stopping assemblage of segments (Figure 8.21). Rotating bearings in the joints simulate the range of human motion. There is no fabric in this suit. It is made entirely of aluminum alloy and stainless steel. The suit weighs about 84 kg (185 lb.) or 36 kg (80 lb.) more than a Shuttle suit.

Vykukal says the suit will be durable, comfortable, easy to inspect after an EVA, and simple to manufacture. In fact, his inspiration was an old wood-burning stove from his childhood home. He and

8.17

The Ames AX-2 experimental hard suit was largely designed by Hubert C. Vykukal. This suit had rotary closures and increased waist flexion range. It featured metal stovepipe joints in the limbs and a six-ply stainless steel bellows at the waist. (SI photo 87-15054)

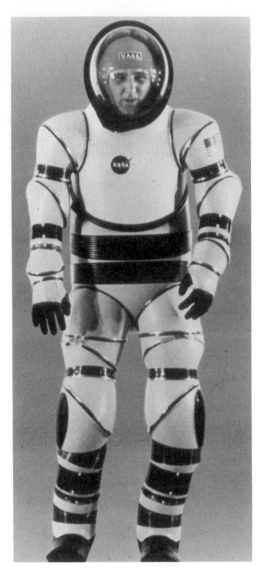

ticulated joints will protect not only the torso and limbs but also the head and feet. The AX-5 would be coated with 70.8 g (2.5 oz.) of gold to prevent corrosion and provide thermal protection. This would protect the astronaut from radiation, micrometeoroids, and man-made debris during EVAs.[9]

Micro Craft, Inc. fabricated the majority of the suit. Air-Lock, Inc. provided the stainless steel bearings, the hatch, connectors, waist sizing rings, and sealed helmets. ILC Dover, Inc. furnished the liquid cooling garment. The cost of the Ames AX-5 program, including two suits, was about $2.5 million.

NASA/JSC has also been developing space suits beyond the RX series hard suits discussed earlier. The Shuttle suit (an outgrowth of the orbital extravehicular space suit), the emergency intravehicular space suit (EIS), and several other suit technology programs of the late 1960s and early 1970s were developed by ILC for the Johnson Space Center.

ZERO-PREBREATHE SUITS

From 1979 to 1983, JSC supported the development of the zero-prebreathe suit (ZPS) concept, components for which were designed and built by several contractors. Design goals for the ZPS included operation at 55 kPa (8 psi), rapid unassisted donning and doffing, quick on-orbit resizing and maintenance, and mobility at least as good as the Shuttle suit provided at 29 kPa (4.3 psi). All of these goals were met or exceeded. The ZPS incorporated innovations as well as designs from earlier advanced suits, even some as

his team of bioengineers took the stovepipe joint and basically improved upon it. Each section is mated to free-moving gimbaled joints with wire and groove rings called Ortman couplings. These sizing rings, 3.5 cm (1.4 in.) long, are also used to increase or decrease ankle-to-knee and knee-to-hip joints. Ar-

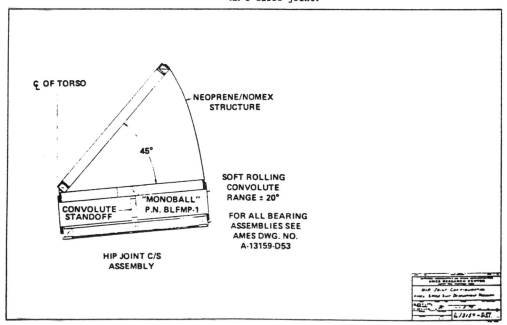

"MONOBALL" P.N.-BLFMP-1

25°
BOTH SOFT
ROLLING
CONVOLUTE
ELEMENTS

CONVOLUTE
STANDOFF

INTERNAL
LINKAGE

70°
WEDGE ELEMENT

NATIONAL AERONAUTICS and SPACE ADMINISTRATION
AMES RESEARCH CENTER

TWO ELEMENT SOFT
ROLLING CONVOLUTE
ELBOW JOINT

NOTED

A-13159-D52A

AX-3 elbow joint.

₵ OF TORSO

NEOPRENE/NOMEX
STRUCTURE

45°

SOFT ROLLING
CONVOLUTE
RANGE ± 20°

CONVOLUTE
STANDOFF

"MONOBALL"
P.N. BLFMP-1

FOR ALL BEARING
ASSEMBLIES SEE
AMES DWG. NO.
A-13159-D53

HIP JOINT C/S
ASSEMBLY

AMES RESEARCH CENTER

A-13159-D53

AX-3 hip joint.

8.18
The AX-3 used a multiple laminate structure of neoprene-coated Nomex and ripstop materials in combination with cable sizing plugs and soft and rolling convolute elements. The AX-3 elbow *(top)* shows both soft and rolling convolute elements; and the hip joint *(bottom)* shows soft rolling convolute. (SI photo 90-8205)

8.19
The AX-3 torso *(top)* incorporated hard structures above and below the dual-plane entry closure and in the section between the waist and hip joint. This allowed for the maximum area to be utilized for mounting the life support system in the rear of the suit. The hip-and-thigh joint featured a sealed bearing *(bottom)* coupled with a soft rolling convolute. (SI photo 90-8203)

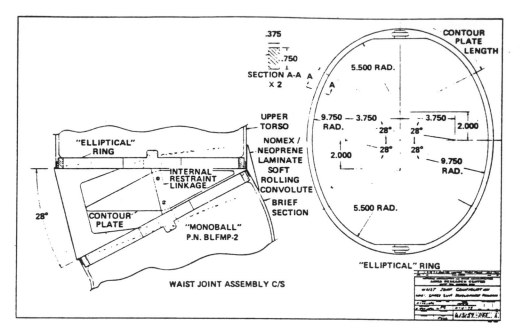

AX-3 waist joint.

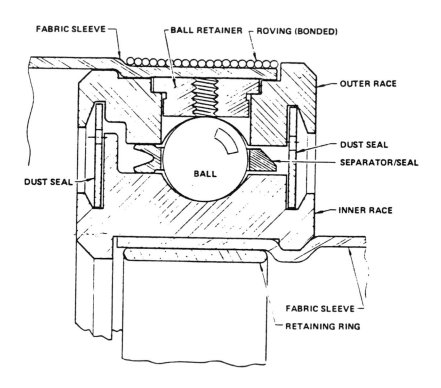

AX-3 sealed bearing (8-in. diam).

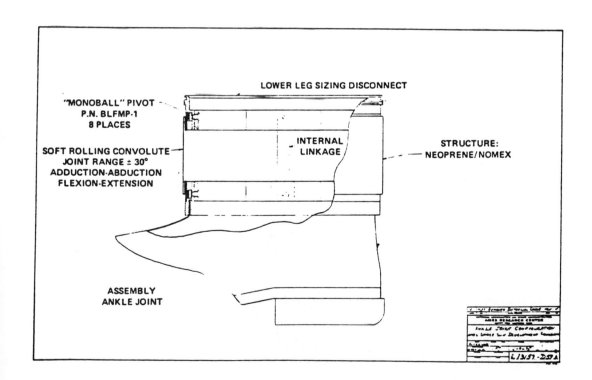

END PLUG /w
JOINT SIZING
DISCONNECT BOTH ENDS
(SEE AMES DWG. A-MOD-D52)

30°

CABLE

60°

CABLE
GUIDES

ALL SOFT STRUCTURE
NEOPRENE/NOMEX
LAMINATE

LOWER LEG SIZING DISCONNECT

"MONOBALL" PIVOT
P.N. BLFMP-1
8 PLACES

INTERNAL
LINKAGE

STRUCTURE:
NEOPRENE/NOMEX

SOFT ROLLING CONVOLUTE
JOINT RANGE ± 30°
ADDUCTION-ABDUCTION
FLEXION-EXTENSION

ASSEMBLY
ANKLE JOINT

8.20

The Ames AX-3 hard suit also incorporated joint sizing disconnects to easily lengthen or shorten limbs as needed. Interchangeable Ortman wire sizing rings varied in width from .76 to 3.18 cm (.3 to 1.25 in.) and were used above and below the knee to increase or decrease length *(top)*. A cutaway of the boot *(bottom)* shows the eliptical opening around the top of the ankle. An Ortman wire sizing ring is fed into the small groove near the area marked for the monoball pivot. (SI photo 90-8204)

8.21
The NASA/Ames AX-5 all-metal rear-entry suit recently underwent evaluation for space station extravehicular activities. Here, it is being tested in the neutral buoyancy test facility by research scientist Hubert ''Vic'' Vykukal. He is testing the suit for comfort and mobility. The device he is grasping helps control movement in the tank. The ballast assembly and midsection control buoyancy. The suit shown is white, but it is planned to have the entire suit coated with a thin layer of gold for reflectivity and corrosion resistance. (SI photo 92-615)

early as Apollo (such as the helmet and glove connectors), plus a Shuttle hard upper torso.[10]

During 1983 and 1984, when NASA was evaluating the ZPS, ILC conducted a company-funded program to design and build an improved ZPS shoulder-and-elbow joint, called the high mobility arm (HMA). This produced a shoulder joint with better mobility and longer life. JSC then started ILC on a second phase to incorporate the HMA shoulder design and other improvements in the ZPS. Toroidal convolute elbows and rolling convolute shoulders in the HMA withstand higher operating pressures but do not constrain motion. HMA joints may contain both fabric and metal rings to maintain flexibility as well as constant volume of gas in the suit. Metal joints move only so far. For the suit knee, a cylindrical joint is constructed of accordion-like fabric folds similar to those at the midsection of articulated buses. Continued force on the joint presses it into the shape of elbow macaroni, and an override capacity extends mobility. If the joint did not expand, the pressure in the suit would increase, making movement difficult. It is a one-way joint. There is not enough material on the underside for the knee to bend the other way. Several other suit technology programs produced all-fabric mobility joints, similar to the Shuttle soft joints, that operated well at 55 kPa and advanced entry/closure designs.

The basic Mark I zero-prebreathe suit mixed a variety of concepts. The ZPS had reinforced pivots in the fiberglass torso and toroidal and rolling convolute joints. New gimbal rings were designed to be the outer race of the shoulder bearing. The bearing consisted of metal inner and outer races connected by several stainless steel balls that ride in the tract. A pressure seal was designed into the bearing race. The metal rings locked together at the waist and helmet disconnects and provided pressure retention through the use of lip seals. The basic ZPS received further upgrades such as additional mobility that provided more reliability for the joint systems. The designation Mark II was used to distinguish this upgraded model from Mark I.

Joe Kosmo, project engineer for the Mark series, says the culmination of these programs was a partly hard, partly soft (because astronauts preferred soft appendages) hybrid: the Mark III ZPS concept (Figures 8.22 and 8.23). JSC built a prototype, along with an enhanced TMG to provide improved protection from radiation, meteoroid debris, and extreme temperature variations. The advanced rear-entry suit was modular so parts could be interchanged to accommodate many users (Figure 8.24). The Mark III is 65-percent metal and weighs about 70 kg (154 lb.) without its thermal overgarment. A variety of surface coatings for thermal protection, such as optical solar reflectors, could be applied to help control surface temperature and heat leaks. Pressurized to about 55 kPa (8 psi), the astronaut would be able to avoid prebreathing and transfer directly from spacecraft to the space environment.

The toughest engineering problem still concerns gloves for the high-pressure suits. Each astronaut will probably carry a wardrobe of three pairs of gloves for each 90-day space stint. There might be one design that is more suitable for manipulative tasks, another for gross handling of large trusses or payload packages.

8.22

Line drawings of the NASA/Johnson Space Center zero-prebreathe Mark III suit concept (1988) show rolling convolute shoulder and the combined use of hard and soft elements. (SI photo 88-1099)

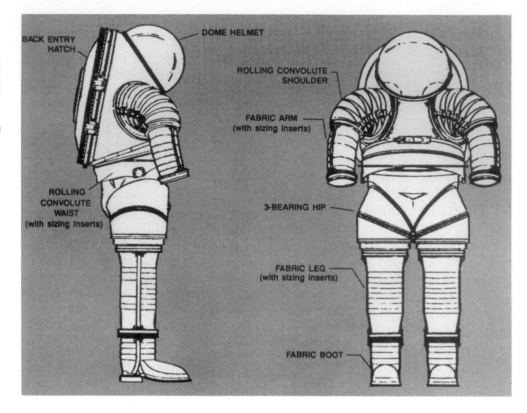

Gloves must not only be protective and flexible, but also durable. Currently, fabrication begins with making casts of an astronaut's hands in one or more positions. The castings are measured by hand in critical areas. Patterns for the glove bladder and restraint layers are defined also by hand. Future gloves may be designed using laser technology to scan an astronaut's hands to form high-fidelity models based on 20,000 bits of data. Engineers will then fit the fabric to the models. As a result of this process, the gloves will be produced in less time and have a more accurate fit.

Final requirements for the space station extravehicular mobility unit have not been established. Present plans call for the Shuttle EMU to be used to construct and maintain the space station. But gradual enhancements of the Shuttle suit continue to be made which eventually could result in a 55-kPa-capability suit with a dual-plane closure as was used in the RX and AX suits.[11]

Designers will continue to refine space suits specifically for living and working in space. Space suits of the future will be worn to perform routine activities outside space stations. They will need to be easy to maintain for at least a year, the probable length of time an astronaut will be stationed in space. NASA outreach offices may well be located in these future space stations. Astronauts could find themselves commuting to space stations just as people commute to their places of business today.

8.23
A technician demonstrates the high flexibility of the NASA/JSC zero-prebreathe Mark III suit. (SI photo 92-643)

8.24

NASA/JSC developed the zero-prebreathe Mark III in conjunction with ILC in Dover, Delaware. First, a liquid cooling garment is put on. Before donning the space suit, the astronaut is connected to a multi-water cooling system via the umbilical. The legs are lowered into the torso through the rear entry hatch and the arms are placed into the front portion of the hard upper torso. The umbilical is connected to the suit hatch. Once inside, the astronaut reaches back and closes the hatch. (SI photo 88-1093)

CHAPTER 9

Space Suits in the National Collection

In 1965, Congress authorized construction of the National Air and Space Museum. Actual collection and accessioning of space suits had begun in 1962 with the receipt of Alan Shepard's training suit. Suits then came into the museum at odd intervals, a few at a time. The increased activity of the manned space programs of Mercury, Gemini, and Apollo caused explosive growth in the collection between 1968 and 1976.

Of course, the most anticipated suits were from *Apollo 11*. Because of tremendous interest and enthusiasm for the space program, the *Apollo 11* spacecraft and space suits toured the fifty states with the astronauts and NASA representatives. Perhaps Senator Ernest Gruening of Alaska explained this phenomenon best. He said the astronaut is among today's most prestigious popular idols. Once launched into space, he holds in his hands something far more costly and precious than the millions of dollars' worth of equipment in his spacecraft: He holds the prestige and the honor of his country. The astronaut is also a symbol of his nation's way of life—and that is what this traveling exhibit represented to the rest of the world.[1]

Completion of the museum became the first priority, followed by the loan program; but sometimes loans took precedence over the new building. Space suits went on loan for

9.2
Environmental archives at the Paul E. Garber Preservation, Restoration, and Storage Facility maintain a steady temperature of 10°C (50°F) and a constant humidity level of about 50 percent. Several of these Bally boxes (named for their manufacturer) have been installed to house special collections such as art, film, rubber tires, and flight materiel.
(SI photo 84-11158-121A)

9.3 *(Facing page)*
Suits and accessories are carefully placed on shelves in metal cabinets inside the environmental storage facility. Each artifact has been identified, inventoried, and tagged.
(SI photo 92-629)

object to determine how it is made, what materials it is made of, the state of condition or deterioration, and possible treatments.

For space suit artifacts, it is a rare decision to choose restoration. Historical evidence such as a tear incurred during the mission, provided it is well documented, cannot be repaired. However, if restoration of the *Gemini 4* suits was not done, their condition would be too fragile for exhibition. Gemini cover layers, constructed from high-temperature-resistant Nomex, were badly discolored and deteriorated around stress areas and seams. The decorative mission patches had faded and were falling off the suit (Figure 9.4). The gold coating on the visor had begun to

peel as well. Ultraviolet (UV) radiation damage due to being on constant exhibit in a sun-lit area had taken its toll.

Options were researched. The David Clark Company, original manufacturer of the Gemini suit, was contacted. Fortunately, patterns for the White and McDivitt space suits had been retained in the NASM archives. The original internal pressure suits would be used with replicated cover layers. This saved the cost of replicating hardware and other space suit materials.

After new cover layers were constructed, the damaged originals were placed in the preservation/study collection in the environmental storage facility. The replicas were used over the original pressure garment assemblies (the support structure of the space suit). But another problem arose. The manikin material originally used had deteriorated into sandy particles fit only for the trash.

In the meantime, two *Apollo 11* suits and one *Apollo 17* suit had been removed for an exhibit upgrade. Apollo cover layers, constructed from Teflon and fiberglass, were more resistant to UV damage. The suits were in excellent condition because they had been on exhibit in a controlled low-light environment. But manikin problems similar to those of the Gemini suits were apparent. Clearly, more appropriate materials and new construction methods had to be found. Information about manikins for museum costumes containing high-technology textiles was scarce. Conservators trained in this new field of plastics were just graduating from universities.

In 1985, Sharon Blank, then an intern at the National Museum of American History's conservation laboratory working in the area of plastics, examined the NASM collections. Among the collections is the pressure suit that Wiley Post wore in record-breaking flights aboard the *Winnie Mae*. Post's suit contains natural rubber. Blank noted that natural rubber, although not a synthetic, is still a high polymer made up of thousands of repeating molecules and subject to many similar problems. She felt it fortunate that the rubber liner was stored in the environmentally controlled room and only the outer shell of the pressure suit was on exhibit.

Blank found the plastics used in space suit parts remarkably diverse, but rapid deterioration of these materials was evident. The latex lining of gloves sometimes crumbled away. Even gloves with the

9.6 (Top left)
Collections Maintenance Chief Al Bachmeier sketched this rough design for an armature to support the manikin wearing a space suit. (SI photo 90-14213)

9.7 (Top right)
These aluminum nodes and modules were available "off the shelf." An exhibits specialist shaped and assembled the armature. (SI photo 89-21486-3)

9.8 (Bottom)
Polyester mesh formed the torso, arms, and legs. In construction of the elbows and knees, a wedge-shaped section of mesh was left out to increase positioning capability. (SI photo 89-21486-7)

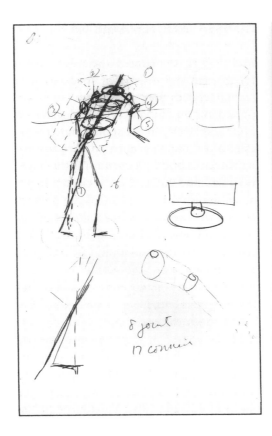

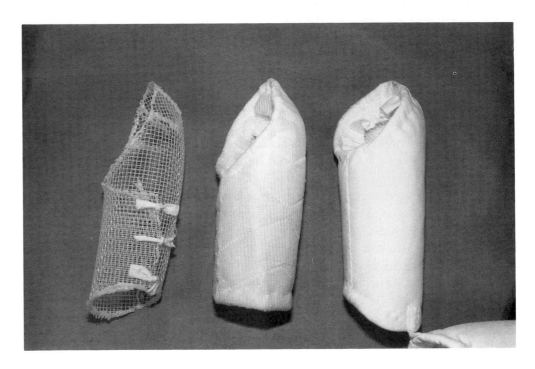

NASM Collections Maintenance Chief Al Bachmeier met with CAL representatives to demonstrate an idea he thought might work and also be practical: an aluminum display system. Looking like an oversize tinker-toy, the armature was constructed of Unistrut display system nodes with pre-drilled holes and long and short aluminum tubing called module lengths (Figures 9.6 and 9.7). Exhibits specialist John Siske devised the method of constructing the pieces into a manikin support structure. Siske bent and shaped the tubing and joined the nodes.

Virginia Pledger, a CAL intern with extensive experience as a textile pattern maker, constructed the manikin torso. She created a pattern that she traced onto nonwoven synthetic material. Using this pattern, she constructed the torso from polyester mesh that she had bent to form

the torso, arms, and legs (Figure 9.8). Once the mesh was cut, a glue coating was applied to cover sharp edges. Knee and elbow areas had a wedge-shaped piece of mesh left out for greater mobility. Mattress padding covered the polyester mesh, followed by cotton sateen for the outermost manikin layer, which provided a protective buffer between space suit and the manikin. Adjustable aluminum bands inside the chest area gave the torso an oval shape. The mesh was formed around the bands. An opening in the rear of the torso gave access for adjusting the armature to make the manikin taller or shorter.

No stuffing was used inside the shoe because the outside materials were stiff enough to hold their shape. Neither *Gemini 4* suit would be exhibited in a standing position, so there would be no weight on the shoes. For Apollo suits exhibited in standing positions, Siske adjusted the armature to suspend the suits just above the floor. The weight of the suit fell on the armature. There would be no strain placed directly on the bottom of the boots.

The manikin for Ed White's suit would have no head since the helmet was displayed with a closed gold visor. The suited manikin held the helmet in place. For the McDivitt manikin, a plain, white molded plaster head had to be cut longitudinally in half to fit inside the helmet's foam liner which had hardened (Figure 9.9). A modified shelf bracket and appropriate couplings held the helmet to the torso. The armature proved strong enough to support Gemini suits weighing 15.4 kg (34 lb.) and the 35-kg (80-lb.) Apollo space suit.

The manikin can be positioned to show an astronaut poised in a spacewalk or seated in an enclosed spacecraft (Figures

9.9
Half of the generic manikin head used for the McDivitt suit had been cut longitudinally to allow for the communication carriers built into the Gemini helmet. The conservator feared that if an intact head were placed inside the helmet, the deteriorating communication earphones, which had already hardened, might make removal of the head impossible. (SI photo 89-21487-6A)

9.10 *(Top)*, **9.11** *(Bottom)*, **and 9.12** *(Right)* The manikins were designed to take various poses, as shown in this display for *Gemini 4*. Astronauts could poise above the spacecraft for a space walk or, because of a rounded modification to the manikin, assume a seated position. (SI photos 89-21483-37A, 89-21487-13A, 89-21484-13)

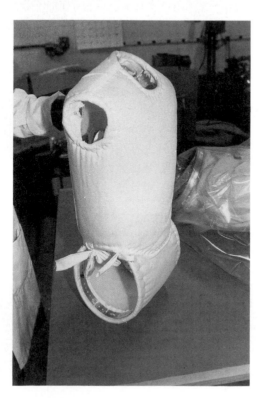

9.10, 9.11, and 9.12). It can be completely taken apart and laid flat for storage, making it appropriate for a traveling exhibit. A condition of loans of space suits from NASM will be the use of this type of manikin.[8]

CONCLUSIONS

Space suits are especially fragile artifacts, easily and often irreparably damaged. They are more susceptible to environmental factors than are many other museum artifacts. Concern for the environment in which space suits are exhibited or stored is fundamental to their preservation. The recognized standards for museum artifacts are stated by Garry Thomson in *The Museum Environment*. The intensity of visible light should measure between 50 and 200 lux.* Ultraviolet radiation should measure less than 75 microwatts per lumen; anything over this measurement requires filtering (Figure 9.13). For example, tungsten lamps read between 60 and 80 lumens, so they usually do not require UV filters. Additionally, artifacts should not be exposed to direct sunlight. Relative humidity should be kept constant, between 40 percent and 55 percent for most materials. Temperature should be maintained between 15.6° and 23.9°C (60° and 75°F).[9]

When the preservation/study collection was examined, new strategies were considered to improve long-term storage. Heat-sealable storage bags constructed of a tough, impermeable film may be assem-

*A lux is equal to the illumination produced by the luminous flux on one lumen incident on a surface with an area of one square meter.

bled under the direction of the conservators. The suits could be stored in an oxygen-free environment with the proper relative humidity. The bags can be constructed with windows so that the suits can be inspected and monitored on a regular basis. Activated charcoal can be used as an oxygen scavenger. This system would particularly address the degradation problems posed by unstable polymers such as the PVC tubing used in the liquid cooling garments.

Dr. Mary Baker, a research organic chemist at the Conservation Analytical

9.13

Light causes damage to museum specimens. All sources of light, whether natural or artificial, emit energy in the ultraviolet and infrared bands of the spectrum in addition to energy in the form of visible light. Plexiglas that provides ultraviolet filtration is given a final polish before installation on the *Gemini 4* exhibit. (SI photo 89-21485-60)

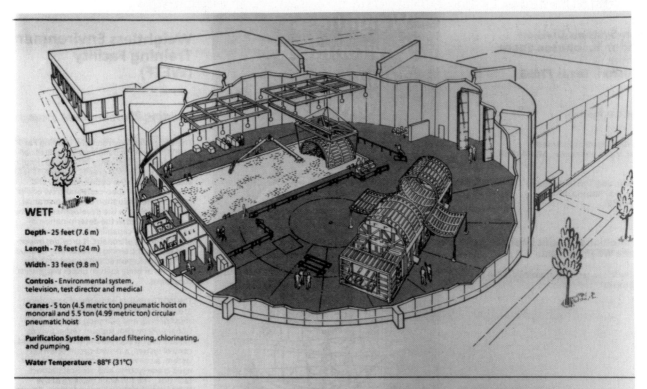

WETF

Depth - 25 feet (7.6 m)

Length - 78 feet (24 m)

Width - 33 feet (9.8 m)

Controls - Environmental system, television, test director and medical

Cranes - 5 ton (4.5 metric ton) pneumatic hoist on monorail and 5.5 ton (4.99 metric ton) circular pneumatic hoist

Purification System - Standard filtering, chlorinating, and pumping

Water Temperature - 88°F (31°C)

Closed-Circuit Television System

A closed-circuit television system is in operation during all pressure-suited test activities. The system consists of two underwater pan-and-tilt units, two underwater hand-held cameras, and test console-mounted television monitors. The hand-held camera is carried and operated by support divers

Dressing Rooms

Dressing rooms for use by male and female test subjects to don and doff the Shuttle Extravehicular Mobility Unit (EMU) are located in Building 29. Separate dressing rooms will be used by the male and female divers.

Medical Station

The station is manned by two trained medical technicians. A medical center doctor is available for immediate response in the event of an emergency. They will monitor the crew activities within the WETF by television. An emergency vehicle is standing-by to transport personnel to the hyperbaric chamber or to the dispensary if there is a

9.14

In NASA's Weightless Environment Training Facility (WETF), closed-circuit televisions are used to monitor all pressure-suit activities. Two trained technicians stand ready at the medical station should an emergency occur. (SI photo 90-9681)

Laboratory, began an in-depth assessment of the stages of space suit degradation in 1990. The Electron Spin Resonance (ESR) test was used to study high populations of free radicals in the glove materials. They were found to be stable. Mercury suits with aluminized nylon coverings showed water discoloration and deterioration. These suits had been tested at the Johnson Space Center Weightless Environment Training Facility (WETF) in deep-water tanks (Figures 9.14 and 9.15).

Baker discovered that the chlorinator used in the tanks was calcium hypochlorite, a strong oxidizer. She tested natural rubber in simulated WETF water and in distilled water. Early results indicate that exposure to the calcium hypochlorite did accelerate the oxidation of rubber. Further testing by Dr. Baker suggests the flaking of the aluminum layers may be due to aging of the adhesive that binds them. It must be determined whether the suits are now stabilized or will degrade further.[10]

9.15
A NASA technician, completely submerged in the WETF, experiences the simulated weightless environment of space, essential training for EVA. (SI photo 90-9085)

Analyses using microspectroscopy techniques have indicated more potential problems from the PVC tubing. As long as it continues to exude plasticizer, it can contaminate other materials in the suits. It can cause softening and stickiness in some polymers and make them more susceptible to common solvents. Instead of storing liquid cooling garments with other space suit materials, perhaps sealing them in oxygen-free plastic bags would prevent further contamination.

Soft rubber parts of gloves, boots, linings, and gaskets also show various stages of degradation. Baker explained that at first the rubber softens and shifts, then hardens into distorted shapes. Adhesives used in laminating aluminum to fabric, as in the case of Mercury and early Gemini and Apollo suits, have become very brittle and the more worn areas have flaked. Foam inside Mercury and Gemini helmets has hardened.

Dr. Baker's initial research centered around the presence of still-reactive materials in the artifacts. This was of concern since storage in an oxygen environment can cause further oxidation and other damage. Still-reactive sites can also cause further cross-linking and, thus, further breakdown in the materials even when stored in an oxygen-free environment, as in the sealed storage bags.

Concern with preservation of high-technology textiles is shared by all museums that house modern textile collections. The National Air and Space Museum now has a full-time conservator on its staff. Conservator Ed McManus's interest and concern for the collection, in conjunction with CAL's research, will further progress. Sharon Blank's observations of the collection had drawn attention to specific problems. Dr. Mary Baker's research using sophisticated laboratory equipment corroborated Blank's theories, but also raised new questions. For instance, will it be cost-effective to store the suits in bags with an oxygen-free atmosphere? Should the entire collection have proper manikins? What restoration is possible for suits in dire need? The study of these and other problems is just beginning.

Conservators should periodically evaluate existing conditions and recommend any necessary improvements. If improper conditions cannot be corrected, space suits should be exhibited in a different manner or not displayed at all. Proper conservation techniques, such as the carefully designed manikins, will certainly extend the life of these unique costumes. Through constant care and vigilance, museum visitors and students will continue to have authentic space suits, and not just replicas, to study and ponder over their origins.[11]

Appendixes

NASA Management Instruction 1052.85
Agreement between the National Aeronautics and Space Administration and the Smithsonian Institution Concerning the Custody and Management of NASA Historical Artifacts
March 3, 1967

In 1967, the National Aeronautics and Space Administration transferred accountability for historic artifacts related to the manned and unmanned spaceflight programs to the Smithsonian Institution. The Smithsonian was then charged with the responsibility of memorializing the national development of spaceflight; collecting, preserving, and displaying aerospace equipment of historical and educational interest and significance; serving as repository for scientific equipment and data pertaining to the development of spaceflight; and providing educational material for the historical study of spaceflight. Through this transfer, the National Air and Space Museum, administered by the Smithsonian, assumed the responsibility for the custody, protection, preservation, and display of all historic artifacts either on its premises or on loans to other educational organizations, including the majority of space suits used in the various developmental stages through manned spaceflight.

APPENDIX 2

The National Air and Space Museum
Guidelines for the Exhibition, Care, and
Handling of Space Suits

Although designed to protect human life in one of the most hostile environments people have ever tried to explore, space suits are not indestructible. They contain both perishable and fragile materials, such as rubber and fabric. Space suits can quickly and irreversibly deteriorate if mishandled or exhibited improperly. The National Air and Space Museum, as the principal repository for the nation's space artifacts, recognizes its obligation to exhibit these objects. Yet the museum also recognizes the need to preserve them for future generations. This effort to preserve the space garments while permitting their exhibition has contributed to the formulation of the following guidelines.

Exhibition Techniques: The preferred method of display is to have the suit supported on a manikin or an aluminum frame which has been padded with 100-percent spun-polyester batting. Care must be taken in the selection of polyester batting because some brands use a starch base, which can attract insects. Wooden frames should be avoided, because wood can release vapors that could damage the suit. Please contact the Department of Space History for detailed information about manikins meeting conservation criteria.

Exhibit Cases: Space suits and other clothing articles are to be housed in secure, protective exhibit cases. Plastic or safety glass should be used to construct the display case. Plate glass, if shattered, produces long, sharp shards that can damage the suit. The case must provide adequate physical protection to prevent theft or damage. Exhibit cases should have vents with changeable filters for dust control and some type of ultraviolet radiation blockage. To reduce heat build-up, lighting fixtures should be located outside the case. The fixtures generate ozone, which accelerates the deterioration of rubber.

Ultraviolet Protection: All wavelengths of light are harmful to museum objects. Synthetic fabrics and rubber are especially susceptible to deterioration from exposure to ultraviolet (UV) radiation. UV light can be particularly damaging when fixtures are close to the artifact. Fluorescent lights, tungsten-halogen lights, and sunlight are three sources of UV radiation. Space suits exhibited in areas where these light sources are encountered should be protected by UV-absorptive materials either as exhibit case walls or as sheathing for fluorescent tubes. We recommend a lumen level no higher than that of 50.

Temperature and Relative Humidity: Temperature should be as low as is acceptable for human comfort, a maximum 21°C (70°F) is recommended. Relative humidity should be maintained at 50 percent ±5 percent. Air conditioning or heating equipment must maintain the proper temperature and relative humidity 24 hours a day and not fluctuate widely. Fabric is moisture-sensitive and will undergo dimensional changes in response to fluctuations in relative humidity. As fabric expands and contracts, dust and dirt particles embedded in the textile can cut through the fibers, weakening and eventually destroying the material.

Traveling Exhibitions: Unless specifically approved, artifacts borrowed from the National Air and Space Museum are not to be used in traveling exhibitions.

Sub-loans: Borrowers are not permitted to loan Smithsonian artifacts they are exhibiting. If an organization or museum wishes to borrow an artifact from the Smithsonian collection on display elsewhere, the object must be returned, and the organization or museum must request and qualify for a loan from the National Air and Space Museum.

Shipping: Suits should be protected from dirt and moisture during shipping. All shipping and packing arrangements are to be approved in advance by the Office of the Registrar, National Air and Space Museum. The Office of the Registrar must be contacted prior to any shipment. Shredded or crumpled newspaper should not be used to pad suits or helmets. All items should be securely packed in well-padded, sturdy wood or metal shipping containers. If metal containers are used, they should not be airtight because of possible damage from condensation.

Cleaning: Borrowers are not allowed to clean space suits. This includes spot cleaning with spray cleaners. Light vacuuming to remove dust is permitted: An upholstery brush covered with cheesecloth (to reduce suction) is recommended.

Storage: Space suits and accessories are generally loaned to be exhibited and should not be stored for extended periods. If an exhibit is undergoing renovation or modification, however, suits may be stored for a brief period. A space suit is best stored unfolded, flat, on a shelf in a dark, secure, clean area. It should not be sealed in a plastic bag as condensation, in certain instances of fluctuating temperature, can form inside the bag and damage the suit. Storing a suit flat keeps the fabric from stretching, as it will if draped across a hanger. It should not be folded, rolled, or compressed, as harm to the rubber and plastic components in the pressure bladder and even the fabric itself may result. If possible, suits and accessories in temporary storage should be covered with material called unbleached cotton print cloth. This is available from Testfabrics, Inc.

Handling: Space suits should not be handled any more than absolutely necessary and only by museum staff. The general public must *never* be allowed to handle a space suit. Exhibitable suits should not be used for hands-on demonstrations or lectures. Curators, exhibit preparators, or other staff members who handle a suit must wear clean, lint-free gloves. Cotton or polyethylene food handler's gloves are acceptable. Space suits, helmets, boots, gloves, coveralls, and other irreplaceable historic artifacts are never to be worn or modeled.

APPENDIX 3

The Materials Revolution— Superpolymers and Plastics

Contrary to popular belief, high-technology synthetic textiles, or the plastics of the space age, were not created expressly for space suits (Figure 1). The word plastic derives from *plastikos,* a Greek word meaning capable of being molded. It is because plastics are deformable, or easily shaped by molding or extruding, that they are so useful. The first plastics were semisynthetic, having a basic chain structure from a natural product such as cellulose. Polymers, fully synthetic compounds, have chains or networks of chemically linked monomers, organic molecules with high molecular weights. Polymerization is the process of forming a polymer from its constituent monomers; this is how plastics are made.

In 1869, John Wesley Hyatt, a printer in Albany, New York, invented **celluloid,** the first commercially successful plastic. Hyatt was seeking a substitute for ivory in one of its popular uses, the billiard ball. He began his research with collodion, a syrupy substance used as a kind of liquid bandage and as a carrier for photosensitive chemicals in a photographic process. Hyatt put ordinary cotton, or some other vegetable fiber, into solution with nitric and sulfuric acids. The end product, nitrocellulose (sometimes called guncotton because of its explosive nature), was then slightly moistened with solvents and mixed under pressure with camphor. This produced a hard, translucent solid that could be formed into spectacular imitations of ivory, tortoise-shell, mother-of-pearl, and amber to make novelties and fancy goods such as buttons, letter openers, hatpins, and jewelry boxes. But one thing celluloid could not do was make a good billiard ball. Its density and elasticity defeated Hyatt's original goal.[1]

Unfortunately, the tendency to consider plastic an "imitation" material rather than a wonderful new substance with properties to be exploited for its own good uses cheapened its image at first. It was Hannibal Goodwin, a Newark, New Jersey pastor, who discovered how to make celluloid into thin, uniform, transparent film, the first non-imitative celluloid product. Henry Reichenback, a chemist at the George Eastman photographic supply house in Rochester, New York, also discovered about this same time, how to produce photographic film from celluloid. However, the pioneers of celluloid did not profit from this product. George Eastman's patents and monopoly practices took care of that.

Rival materials began to appear on the market as chemists explored the impressive structure of polymers. Leo Baekeland, a Belgian-trained chemist, developed the synthetic resin **Bakelite** in 1907. Hyatt's plastic company quickly began producing quality

This appendix was originally presented as a paper titled "Billiard Balls to Moon Suits: The Materials Explosion" for the Ars textrina Eighth Annual Conference on Textiles at the University of Wisconsin, Madison, Wisconsin, June 22–24, 1991.

billiard balls from Bakelite. Because of the flammability of celluloid, **cellulose acetate,** a somewhat more expensive but safe film, was developed. A distant cousin, **cellophane** material, was discovered in 1908. Celluloid products continued to be important into the 1930s, when most of the old companies were absorbed by manufacturers such as Celanese and Du Pont.

In 1926, Charles Stine, director of E. I. du Pont de Nemours, Inc., central chemical research department, decided to argue for a pure science program. He wanted to discover new scientific facts as opposed to applying previously established information to practical problems. Wallace Hume Carothers (Figure 2), a 31-year-old Ph.D. from the University of Illinois, was an instructor at Harvard when Stine recruited him. In his nine years with the Du Pont Company, Carothers made many important contributions to polymer science. He also led the research efforts that produced **neoprene** synthetic rubber and **nylon.**[2]

Dr. Carothers began the first chemical synthesis of fibers from nonfibrous materials for the Du Pont Company in 1928. **Polyisoprene,** a substance with distinct rubberlike qualities, had been first produced in 1880. Du Pont researchers had tried unsuccessfully to create synthetic rubber from divinylacetylene (DVA), a short polymer consisting of three acetylene molecules. Carothers and fellow chemist Arnold Collins prepared to polymerize very pure DVA. In the process, they decided to isolate and identify impurities that tended to yellow their preparation. After distilling the crude DVA, they found a new liquid that solidified to a white, rubberlike mass that sprang back to its original shape when deformed. This was the first sample of neoprene.

Neoprene, made from the polymerization of chloroprene, is characterized by its resistance to substances that contain oil. Neoprene synthetic rubber had first revolutionized the world of manufacturing in 1932 with hundreds of new industrial applications and durability greater than natural rubber. However, it was the demands of World War II, with Japan controlling the rubber plantations of Southeast Asia, that pivoted the production of rubber into a special synthetic rubber project.

Carothers was soon joined by Dr. Julian Hill (Figure 3). Hill discovered a synthetic fiber while attempting to produce polymeric chains longer than anyone had ever prepared in the laboratory. The two scientists had been thwarted in their progress because water, which formed as a by-product, limited the molecular chain growth. Carothers and Hill built a molecular "still" to remove this excess water. They began heating an unusual acid-alcohol combination because Carothers had decided that the reaction of a 16-carbon-chain acid with a short 3-carbon-chain alcohol would promote the formation of the desired longer molecules. As they removed a sample of the resultant product, Hill observed the molten polymer could be drawn into fibers. Once the fibers cooled, they could be further stretched. By 1931, they reported their early findings to the American Chemical Society in a paper titled "Artificial Fibers from Synthetic Linear Condensation Polymers." Du Pont was already involved in the production of rayon products, when the pair of chemists worked out refining processes for forming fibers from polymers. The basic raw materials were hydrocarbon (from coal, petroleum, or natural gas), nitrogen and oxygen (from air), and hydrogen (from water).

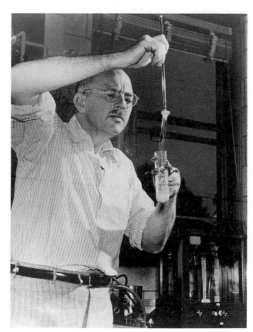

2 *(Left)*
Dr. Wallace H. Carothers joined the Du Pont Company in 1928 to lead a team of researchers in fundamental organic chemistry. His work led to the discovery of neoprene, the first successful synthetic rubber, and nylon, the first man-made polymer. (photo courtesy of the Du Pont Company, SI photo 91-1938)

3
Dr. Julian Hill pulls a molten mass of polymer out of a test tube, stretching it into a thin fiber. After cooling, the polymer could be extended into very long fibers. Further refinements led to the development of nylon. (photo courtesy of the Du Pont Company, SI photo 91-1936)

This first material was called rayon 66 or simply polymer 66; the first 6 indicated the number of carbon atoms in the basic chain, the second 6 indicated the number of acid radicals (NH_3). Technicians devised methods of extruding the fiber, winding it on bobbins, finishing it, and knitting it into stockings on machines originally intended for silk.

There were many disappointments and frustrations. Spinning out strands of fiber from a melted mass required special spinnerets, through which the melted polymer could be extruded. The first stockings produced did not fit standard forms previously used for silk. Because nylon* is thermoplastic (changing shape when exposed to heat), technicians had to work out heat-setting treatments. It took $27 million and 11 years of research and development to make nylon commercially successful (Figure 4). A decision was made not to register nylon as a trademark but rather keep it a generic term to cover all types of nylon such as bristles, sheets, and other forms from the polyamide family of compounds.

National sales of nylon stockings began in early 1940. Hardly had the American appetite for nylons been whetted when the attack on Pearl Harbor embroiled the United States in World War II. Production of the gossamer hosiery yarns was ceased in

---

*The word nylon was derived from one of its early uses: a replacement for silk stockings. A naming committee started with "no run" and ended up with "nylon," much easier to remember than "hexadecamethylene dicarboxylics."

4 *(Top)*
The public first viewed nylon stockings on models at the New York 1939 World's Fair. (photo courtesy of the Du Pont Company, SI photo 91-1927)

5
In the first post-war sale of nylons in San Francisco in 1945, an estimated 10,000 shoppers fill Market Street to get into a store that advertised the precious hosiery. This sale was called off after one store window was pushed in by the force of the crowd and several women fainted. (photo courtesy of the Du Pont Company, SI photo 91-1931)

favor of heavier yarn for parachutes, flight clothing, ropes, and tires. Because of nylon's strength and resistance to moisture and mildew, the entire output was allocated for vital war needs. Even old nylons would be melted down and reprocessed into parachute cloth. Following the war, nylon stockings were once again available and women punched each other and broke windows to get them. The nylon riots were considered the peak of the "golden age of plastics" in America (Figure 5). Nylon safety netting has since protected the Space Shuttle and a nylon climbing rope attached to a harness is now part of the crew escape system.

Dr. Carothers also laid the scientific foundation for the development of polyester fibers. He synthesized several polyesters in the laboratory, but found them disappointingly inferior to nylon polymers because of their low melting point. J. R. Whinfield and J. T. Dickson of Great Britain carried Carothers's work a few steps further and are credited with the development of the first commercial polyester fibers. In the United States, the Du Pont Company purchased Whinfield and Dickson's patent and subsequently obtained a U.S. patent for it. Small-scale experimental production began in 1949 of Fiber V. The temporary name of this new fiber was later dropped for the trademark name of **Dacron.** Following the war, when Du Pont research again turned to peacetime needs, Dacron polyester was introduced commercially in 1950, shocking the traditional male bastions of fashion with wash-and-wear suits and shirts and liberating women from their ironing boards.

Teflon was first discovered in 1938 by Roy Plunkett (Figure 6) at Du Pont while researching refrigeration gases. Plunkett had connected his cylinder of Freon 1114, or tetrafluoroethylene, to a container. He later identified a thin waxy residue at the bottom of the can as a polymer of tetrafluoroethylene or TFE. As with nylon, it took many years of research and much investment of money to understand Teflon's chemical properties. It was used only in limited ways for critical defense projects during the war and became more widely used in the late 1950s. As in all polymers, Teflon is based on a chain of carbon atoms, but unlike others, its chain is completely surrounded and protected by fluorine atoms, making their bond particularly strong. This unique construction gives Teflon such extraordinary properties as its inertness to almost all chemicals and a temperature range from $-240°$ to $260°C$ ($-400°$ to $500°F$). Teflon is widely used in printed circuitry and insulation.

In 1962, Frederic S. Dawn was studying textiles at the Manned Spaceflight Center in Houston, Texas, following the *Apollo 1* fire. He looked specifically for non-flammable material that would not support combustion in the pure-oxygen atmosphere of the Apollo spacecraft. That was why he decided to research in the man-made polymeric area rather than in organic-based textiles. NASA sought a flexible, lightweight, strong, non-flammable material for their Apollo space suit. Dawn discovered that spun glass manufactured by the Dow-Corning Company under the trade name of **Fiberglas** had possibilities. He began working with Dow-Corning researchers at their Culbertson, Rhode Island, research laboratory on material technically called **Beta silica** fiber, a different material than that used in Fiberglas.

Fiberglas in fibrous form is used in making various products such as glass wool, yarns, and textiles. Beta silica fiber could be spun into thread fine enough so that

6
Roy Plunkett (*right*) discovered Teflon, which is inert to almost all chemicals and has an extremely wide temperature range. (photo courtesy of the Du Pont Company, SI photo 91-1935)

7 (*Facing page*)
Du Pont developed 20 of the materials used in the Apollo space suit. None of the materials were developed specifically for the space program. Some were new materials, such as Kapton, while others, such as nylon, had been on the market for years. (illustration courtesy of the Du Pont Company, SI photo 87-7719)

materials woven from it would not irritate the wearer. For some items of space wear, Beta cloth that was not coated was manufactured. For other items, a super Beta cloth was manufactured: Beta silica fiber coated with Teflon. Super Beta cloth was used as the cover layer of the Apollo space suits. Beta silica fiber, one of the few non-Du Pont materials used in the Apollo space suit (Figure 7), is also used to form huge canopies over football stadiums and other large structures.[3]

Another non-Du Pont product used inside the spacecraft as well as on space suits is **Velcro.** In 1948, Swiss engineer George de Mestral puzzled over cockleburs covering his clothing after a walk in the woods. He examined them under a microscope and found

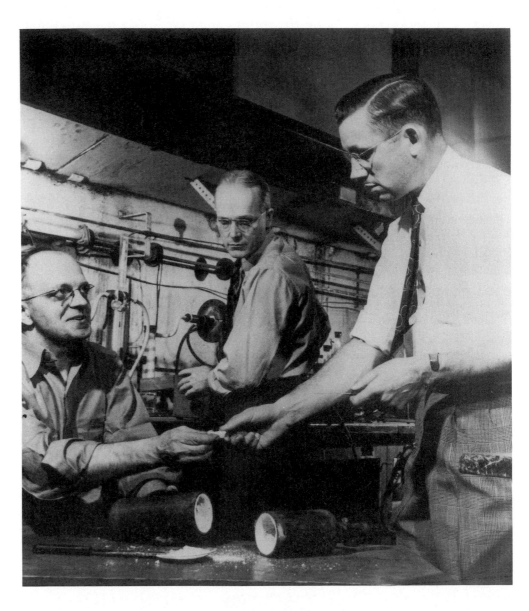

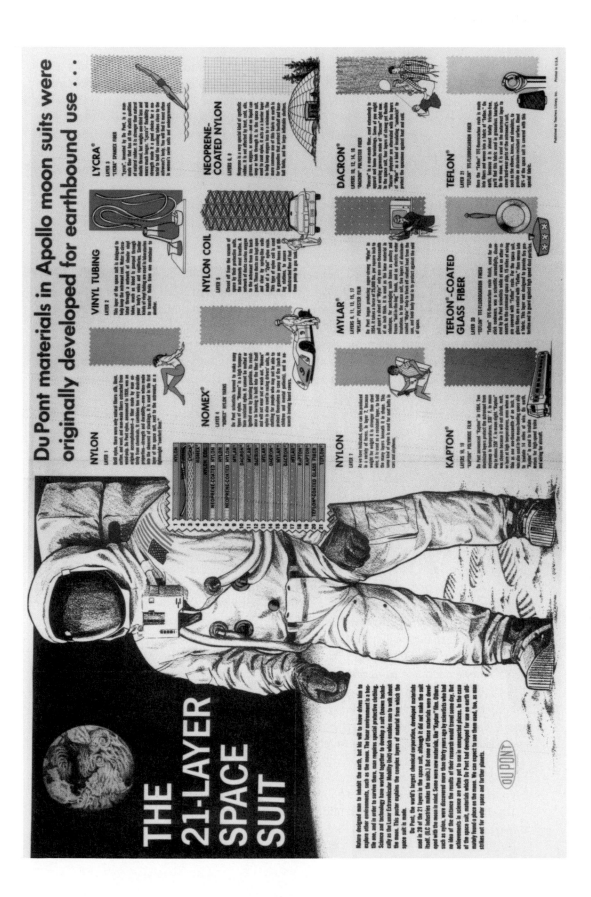

that they were covered with tiny hooks that snagged anything fuzzy. He thought a man-made adhesive could be developed on this same principle. Of course, it took many years of experimentation before it was finally marketed in 1957 as Velcro (*vel* for velvet and *cro* for hook, as in crochet). De Mestral needed to produce a fabric that burs would stick to reliably. He found that nylon thread woven in loops and then heat-treated could grip burs and retain its shape and resiliency over long periods. With the help of a loom maker, de Mestral devised a suitable manufacturing process and Velcro was born.

Unfortunately, there seemed to be little market for such a novelty at first. Then came the manned space program. Velcro fasteners proved ideal for keeping small objects, such as writing pens and notepads in place. **Astro Velcro,** fashioned from Teflon loops, polyester hooks, and Beta glass backing, even allowed astronauts to scratch itchy noses on patches mounted inside their helmets. Astro Velcro holds down the cargo blankets in the Shuttle bay. It keeps lunch from wandering around. It holds astronauts in their workplace and prevents sleeping astronauts from taking microgravity sleepwalks. Each Shuttle contains 65,000 cm² (10,000 in.²) of Astro Velcro.

Velcro is used in medicine as well. For instance, the two pumping chambers in the Jarvik artificial heart are attached with Velcro. If one fails (as happened with pioneer patient Barney Clark), it may be replaced without removing the entire unit. Silent Velcro is also on the horizon. The U.S. Army is replacing snaps on pocket closures of uniforms with Velcro. They worried at first that the distinctive rip could betray soldiers trying to conceal their position. But Velcro researchers have been able to quiet 95 percent of the noise.[4]

Production of synthetic polymers skyrocketed. During the 1950s, the transition was made from using plastics to imitate natural products to appreciating plastics for their own properties. The market was filled with products such as Tupperware, Hula-hoops, inflatable furniture and buildings, wigs, and dishes. Back at the Du Pont Company in the mid-1950s, chemists produced **Mylar** polyester film as one of the first materials to be used for atmospheric research balloons. Mylar was also used in overhead projection transparencies and later became drafting and engineering material for blueprints. It also became a superior base for microfilm and magnetic recording tape. In 1959, Lycra **spandex** fiber brought stretch-and-recovery quality to nylon swimwear, lingerie, and denim jeans. Another outgrowth in 1963 of nylon research produced **Nomex,** of the aramid polymer family. The high flame-resistance of Nomex made it especially beneficial to protect fire fighters and professional racing-car drivers. Mercury and Gemini space suits were made from this material.

In 1969, W. L. Gore & Associates discovered that stretching Teflon into a superthin membrane created a breathable, waterproof material useful for cold- and wet-weather wear, very popular with sports enthusiasts. **Gore-Tex** ortho-fabric woven materials were later chosen for the cover layer of the Shuttle suit. The Shuttle suit uses a total of 177.5 m (195 yd.) of fabric. Of this, 21 m (23 yd.) are Gore-Tex ortho-fabric, 15.5 m (17 yd.) are urethane-coated nylon, 16 m (18 yds.) are Dacron, 20 m (22 yd.) are Neoprene-coated ripstop, and 105 m (115 yd.) are aluminized reinforced Mylar.[5]

In 1965, Stephanie Kwolek and Paul Morgan, research scientists at the Du Pont experimental station, were puzzled about an opaque liquid that Kwolek could clear by heating or filtration. This usually meant that some inert matter was dispersed in the spin dope and was plugging the spinneret holes. Kwolek extruded the liquid through the spinneret anyhow. It spun. Further research and development refined the opaque substance, turning up a polymer that eventually became the basis for **Kevlar.** Herbert Blades, research scientist, went against conformity and heated the polymer solution containing pure sulfuric acid. This allowed spinning at much higher polymer concentrations. He also devised an air gap spinning technology that ultimately lead to the present Kevlar. The entire process took 40 manpower years of development.

Kevlar was ready for the market in 1972. Kevlar is five times stronger than steel. The Gore-Tex cover layer material for the Shuttle suit is reinforced by Kevlar. It is successfully used as well for belting on tires. Lightweight composites for modern civilian and military aircraft are reinforced with Kevlar. Law enforcement officers' lives have been saved by Kevlar "body suits."[6]

Other polymeric elastomers (elastic substances resembling rubber with a chemical composition made up of many long-chained, repeated structural units, such as Teflon), like **Kapton** polyimide film in the 1970s, greatly expanded the use of superpolymers in transportation and for high-temperature seals, wire wraps, and insulation in motors. Protective clothing for pilots, and later for astronauts, incorporated inventive man-made materials as they appeared on the market, from nylon in the 1940s through today's newest fabrics.[7]

NOTES

1. Robert Friedel, "The First Plastic," in *American Heritage of Invention and Technology,* Summer 1987, vol 3, no. 1, pp. 18–22.

2. David A. Hounshell and John Kenly Smith, Jr., "The Nylon Drama," in *American Heritage of Invention and Technology,* Fall 1988, vol 4, no. 2, pp. 40–49. Kathryn A. Jakes, "The Development of Modern Polymers," and Lisbeth G. Stockman, "Polyester: History, Characteristics and Preservation," 9th Symposium, "20th Century Materials, Testing and Textile Conservation," November 3–4, 1988, Harpers Ferry Regional Textile Group, Washington, D.C.

3. Telephone conversation with Frederic S. Dawn, Johnson Space Center, April 19, 1989.

4. Dennis Holder, "Sticky Business," *Philip Morris,* Fall 1988, vol 3, no. 4, pp. 27–29. John Flinn, "Manchester's Velcro Celebrates 40-Years of Stick-To-Itiveness, Fastener's Rrrrrr's can be Heard Everywhere," *The Union Leader* (Manchester, NH), December 9, 1983, n.p. I am grateful to Katherine Dirks, Textile Conservator, National Museum of American History, Smithsonian Institution, for bringing these articles to my attention.

5. Ann Davis Nunemacher, "Layer Upon Protective Layer Seamstresses Practice Spaceage Sewing for NASA," *Sew News,* December 1986, n.p.

6. D. Tanner, J. A. Fitzgerald, and B. R. Phillips, "From Laboratory to Marketplace Through Innovation," booklet based on a paper presented at the Du Pont Company Advanced Materials Conference, October 25, 1988.

7. "Nylon's Golden Anniversary," Background News, E. I. du Pont de Nemours & Company, Inc., Wilmington, Delaware, January 1988, material made available from Du Pont's Public Relations Department, prepared especially for the golden anniversary celebration. Roger Bruns, "Of Miracles and Molecules, The Story of Nylon," in *American History Illustrated,* vol. XXIII, no. 8, December 1988, pp. 25–29, 48. Henry Allen, "Their Stocking Feat, Nylon at 50 and the Age of Plastic," *The Washington Post,* January 13, 1988, pp. D1, 11.

APPENDIX 4

Pressure Suits in the Aeronautical Collection
of the Preservation/Study Collections
Paul E. Garber Preservation, Restoration, and Storage Facility

Model	Suit Type	Fund	Year	Manufacturer	Cost/Source (dollars)
W. Post	Full pressure	Private	1934	Goodrich	75/Colley
S-2	Partial pressure	USAF	1951	USC/Clark	450/Schulz
S-4	Partial pressure	USAF	1952	Clark	450/Schulz
T-1	Partial pressure	USAF	1953	Clark	1,500/Mallan
MC-1	Partial pressure	USAF	1954	Clark	3,000/USAF
MC-3	Partial pressure	USAF	1956	Clark	769/USAF
MC-3A	Partial pressure	USAF	1956	Clark	800/USAF
MC-4[a]	Partial pressure	USAF	1957	Clark	850/USAF
MC-4A	Partial pressure	USAF	1960	Clark	900/USAF
CSU/3P	Partial pressure	USAF	1956	Clark	2,600/USAF
CSU/4P	Partial pressure	USAF	1957	Clark	2,666/USAF
CSU/5P	Partial pressure	USAF	1958	Clark	3,000/USAF
Mark II[b]	Full pressure	USN	1956	Goodrich	[c]
Mark IV, Model 2, Type 1A	Full pressure	USN	1960	Goodrich	[c]
Mark IV, Type 2B	Full pressure	USN	1960	Goodrich	[c]
AP22S-3 (modified Navy Mk IV)	Full pressure	USAF	1960	Goodrich	[c]
CSK-6/P-A/P22S-2	Full pressure	USAF	1964	Clark	3,891[d]
CSK-8/P22S-4	Full pressure	USAF	1966	Clark	4,000[d]
CSK-9/P22S-6	Full pressure	USAF	1968	Clark	4,000[d]
Passive (Boyle's Law) (modified Mk IV)	Full pressure	USN	1968	Goodrich	[c]
S-998 (X-15 type)	Full pressure	USAF	1968	Clark	4,000[d]
S-901E (SR-71 type)	Full pressure	USAF	1968	Clark	4,000[d]
S-901J[b]	Full pressure	USAF	1973	Clark	4,000[d]

[a]With anti-g protection. Partial pressure suits require no aircraft modification, while full pressure suits can only be used in aircraft modified to provide sufficient ventilation.

[b]Similar to type worn on Shuttle flights for five launch and reentries.

[c]Part of $125,000 purchase order to cover research and development for five Mercury suits (NASA/Mercury Chronology).

[d]A. F. Physiological Training Program Newsletter, No. 44, June, 1961. Suits were all improved modifications of the AP22S-2.

Note: The Preservation/Study Collection was established to preserve representative stages of the development of pressure garments. The collection is stored at the Paul E. Garber Restoration Facility in Suitland, Maryland, and is open by reservation to qualified researchers.

APPENDIX 5

Pressure Suits in the Space Collection
of the Preservation/Study Collections
Paul E. Garber Preservation, Restoration, and Storage Facility

Model	Suit Type	Fund	Year	Manufacturer	Cost (dollars)
Mark IV	Full pressure	USN	1960	Goodrich	3,250[c]
Mark V	Full pressure	USN	1968	Goodrich	3,250[c]
Mercury	Full pressure	USN	1961–63	Goodrich	10,000[c]
Gemini G2G	Full pressure	USN	1962	Goodrich	[d]
Gemini G1C	Full pressure	USAF	1962	Clark	8,588[e]
Gemini G2C	Full pressure	USAF	1962	Clark	8,588
Gemini G3C	Full pressure	NASA	1965	Clark	16,000[f]
Gemini G4C	Full pressure	NASA	1965	Clark	80,000[c]
Gemini G5C	Full pressure	NASA	1965	Clark	16,000
Gemini S1C	Full pressure	NASA	1966	Clark	80,000[g]
Apollo A1C	Full pressure	NASA	1964–67	Clark	16,000[h]
Apollo A4H	Full pressure	NASA	1964	HS/ILC	44,000[i]
Apollo A5H	Full pressure	NASA	1964	HS/ILC	54,000[i]
Apollo A2L	Full pressure	ILC	1963	ILC	54,000[i]
Apollo AX5L	Full Pressure	ILC	1964	ILC	54,000[i]
Apollo A5L	Full pressure	ILC	1964	ILC	54,000[i]
Apollo A6L	Full pressure	NASA	1967	ILC	54,000[i]
Apollo A7L	Full pressure	NASA	1968–71	ILC	54,000[i]
Apollo A7LB	Full pressure	NASA	1971–72	ILC	54,000[i]
MOL	Full pressure	USAF	1964	HS	54,000[i]
Skylab A7LB	Full pressure	NASA	1972	ILC	54,000[i]
ASTP[a]	Full pressure	NASA	1975	ILC	54,000[i]
Shuttle eject[b]	Full pressure	NASA	1984	Clark	5,000[j]
RX-2 hard suit	Full pressure	NASA	1964	Litton	75,000[j]
RX-3 hard suit	Full pressure	NASA	1966	Litton	960,000
RX-4 hard suit	Full pressure	NASA	1967	Litton	991,000
AES	Full pressure	NASA	1967	Litton	1,319,310

[a]Modified A7L.

[b]Like S-901J.

[c]*Suiting Up,* Mallan, p. 191, p. 221.

[d]NASA News Release, MSC 63-110, 7/5/63.

[e]Project Gemini Chron, p. 37.

[f]Missiles and Rockets, 5/25/65, p. 31.

[g]Experimental lunar suit.

[h]A1C, which was just like the G3C, was changed following the *Apollo 1* fire to more heat-resistant materials.

[i]Apollo space suit development memo, NASM archives, 3/64. The total contract between Hamilton Standard and ILC to develop Apollo suits was for $62 million.

[j]Handwritten notes by John Thiel (MSC) in NASM archives. Total spent was $8 million on hard suit development.

APPENDIX 6
Project Mercury Summary

Date of Launch	Date of Landing	Flight	Pilot*	Suit	Spacecraft	Objectives/ Accomplishments	Flight Duration hr:min:sec
5/5/61		MR-3	Shepard	M-22	Freedom-7	Suborbital; first American in space	00:15:22
7/21/61		MR-4	Grissom	M-21	Liberty Bell-7	Suborbital	00:15:37
2/20/62		MR-6	Glenn	M-23	Friendship-7	3-orbit flight; first American in orbit	04:55:23
5/24/62		MA-7	Carpenter	M-26	Aurora-7	3-orbit flight	04:56:05
10/3/62		MA-8	Schirra	M-15	Sigma-7	6-orbit flight	09:13:11
5/15/63	5/16/63	MA-9	Cooper	M-20	Faith-7	22-orbit flight to evaluate effects on humans in space	34:19:49

*All were one-person crews.

Note: Space suits listed in these appendixes are not available for loan. The space suit loan program has been very popular since its inception, thus the majority of suits are already on loan to other institutions.

APPENDIX 7
Project Gemini Summary

Date of Launch	Date of Landing	Flight	Crew	Suit	EVA Date	EVA Duration hr:min:sec	Objectives/ Accomplishments	Flight Duration hr:min:sec
3/23/65		GT-3	Grissom (CP) Young (P)	G3C-1 G3C-4			First 2-man mission	04:53:00
6/3/65	6/7/65	GT-4	McDivitt (CP) White (P)*	G4C-3 G4C-8(v)	6/3	00:20:00	First American EVA	97:56:11 (4 days)
8/21/65	8/29/65	GT-5	Cooper (CP) Conrad (P)	G4C-10 G4C-15			Showed feasibility of lunar landing	190:55:14 (7 days)
10/25/65		GT-6					Agena lost; mission aborted	
12/4/65	12/18/65	GT-7	Borman (CP) Lovell (P)	G5C-5 G5C-6			Tests	330:35:31 (13 days)
12/15/65	12/16/65	GT-6A	Schirra (CP) Stafford (P)	G3C-3 G4C-21			Rendezvous with GT-7	25:51:24 (1 day)
3/16/66		GT-8	Armstrong (CP) Scott (P)	G4C-24 G4C-27(v)			Mission curtailed	10:41:26
5/17/66		GT-9					Target vehicle failed; mission aborted	
6/3/66	6/6/66	GT-9A	Stafford (CP) Cernan* (P)	G4C-17 G4C-32(v)	6/5	02:07:00	AMU test	72:21:00 (3 days)
7/18/66	7/21/66	GT-10	Young (CP) Collins* (P)	G4C-19 G4C-36(v)	7/19 7/20	00:40:00 00:39:00	Science; SEVA photo; retrieve experiment	70:46:39 (2 days)
9/12/66	9/15/66	GT-11	Conrad (CP) Gordon* (P)	G4C-39 G4C-40(v)	9/14	00:33:00 02:05:00	Science; SEVA photo	71:17:08 (2 days)
11/11/66	11/15/66	GT-12	Lovell (CP) Aldrin* (P)	G4C-41 G4C-42(v)	11/11 11/12 11/13	00:55:00 02:06:00 00:55:00	Science; photography; photography	94:34:31 (3 days)

*EVA astronaut

Note: A suit modification to the basic G4C suit (a variant) is denoted by (v).

Project Apollo Summary

Date of Launch	Date of Landing	Flight	Crew	Suit	Spacecraft	Flight/EVA Duration hr:min:sec	Objectives/ Accomplishments
10/11/68	10/22/68	Ap-7	Schirra (CDR) Eisele (CMP) Cunningham (LMP)	A7L-004 A7L-005 A7L-006	CM-101	260:09:03 (10 days)	First Apollo mission
12/21/68	12/27/68	Ap-8	Borman (CDR) Anders (CMP) Lovell (LMP)	A7L-030 A7L-031 A7L-037	CM-103	147:00:42 (6 days)	First flight to moon
3/3/69	3/13/69	Ap-9	McDivitt (CDR) Scott (CMP)* (SEVA) Schweickart* (LMP)	A7L-020 A7L-019 A7L-015	CM-104, Gumdrop LM-3, Spider	241:00:54 (10 days) EVA 3/7, 00:37:00, SEVA	EVA from LM; film and thermal samples retrieved
5/18/69	5/26/69	Ap-10	Stafford (CDR) Young (CMP) Cernan (LMP)	A7L-047 A7L-043 A7L-044	CM-106, Charlie Brown LM-4, Snoopy	192:03:23 (8 days)	First LM orbit of moon
7/16/69	7/24/69	Ap-11	Armstrong* (CDR) Aldrin* (LMP) Collins (CMP)	A7L-056 A7L-077 A7L-033	CM-Columbia LM-Eagle	195:18:35 (8 days) EVA 7/21, 02:31:40	First lunar landing; dual lunar EVA
11/14/69	11/24/69	Ap-12	Conrad* (CDR) Bean* (LMP) Gordon (CMP)	A7L-065 A7L-067 A7L-066	CM-Yankee Clipper LM-Intrepid	244:36:25 (10 days) EVA:11/23, 03:56:00; 11/24, 03:49:00	Lunar landing Scientific experiments
4/11/70	4/17/70	Ap-13	Lovell (CDR) Haise (LMP) Swigert (CMP)	A7L-078 A7L-061 A7L-088	CM-Odyssey LM-Aquarius	142:54:41 (6 days)	Lunar landing aborted
1/31/71	2/9/71	Ap-14	Shepard* (CDR) Mitchell* (LMP) Roosa (CMP)	A7L-090 A7L-073 A7L-085	CM-Kitty Hawk LM-Antares	216:01:57 (9 days), EVA 2/4, 04:49:00; 2/5, 04:35:00	Lunar landing Photos; exploration

Date	Mission	Crew	Suit	Spacecraft	Duration / EVAs	Notes
7/26/71	Ap-15	Scott* (CDR) Irwin* (LMP)	A7LB-315 A7LB-320	CM-Endeavor LM-Falcon	295:11:53 (12 days) Scott, SEVA, 7/30, 00:33:00; EVAs: Scott, Irwin, 7/31, 06:33:00;	First use of LR Photos
		Worden* (CMP)	A7LB-094	LR-1	8/1, 07:12:00; 8/2, 04:50:00; Worden, trans-earth EVA, 00:38:00	Lunar exploration Science
4/16/72	Ap-16	Young* (CDR) Duke* (LMP)	A7LB-322 A7LB-327	CM-Casper LM-Orion	265:51:05 (11 days) EVAs: Young, Duke 4/21, 07:11:00; 4/22, 07:23:00; 4/23, 03:40:00	Filming Science
		Mattingly* (CMP)	A7LB-082	LR-2	Mattingly, 01:24:00 trans-earth EVA	Drill lunar core samples; science; drill samples
12/7/72	Ap-17	Cernan* (CDR)	A7LB-328	LM-Challenger	301:51:59 (12 days)	Gravity experiment
12/19/72		Schmitt* (LMP)	A7LB-329	CM-America	EVAs: Cernan, Schmitt, 12/11, 07:12:00; 12/13, 07:37:00; 12/13 07:15:00	Lunar exploration
		Evans* (CMP)	A7LB-404	LR-3	12/13, 07:15:00 Evans; trans-earth EVA, 01:06:00	Exploration

*EVA astronaut

APPENDIX 9
Project Skylab Summary

Date of Launch	Date of Landing	Flight	Crew	Suit	EVA Date, hr:min:sec	Mission Duration days:hr:min:sec	Objectives/ Accomplishments
5/14/73	7/11/79	SL1	unmanned			6 years	100-ton space station; 3 crews visited; 34,981 orbits
5/25/72	6/22/72	SL2	Conrad (CDR)	A7LB-614	5/25, 00:37:00	28:00:49:49	Weitz: repair boom;
			Weitz (PLT)	A7LB-616	6/7, 03:30:00		Kerwin, Conrad: repair boom;
			Kerwin (SPT)	A7LB-615	6/19, 01:36:49		Weitz: parasol deploy
7/28/73	9/25/73	SL3	Bean (CDR)	A7LB-632	8/6, 06:30:00	59:11:09:04	Garriott, Lousma: ATM care
			Lousma (PLT)	A7LB-634			
			Garriott (SPT)	A7LB-633	8/24, 04:30:00 9/22, 02:45:09		Bean, Lousma: add new gyroscopes; Garriott: retrieve 9 experiments
11/16/73	2/8/74	SL4	Carr (CDR)	A7LB-626	11/23, 06:33:00	84:01:15:31	Pogue, Gibson: ATM care;
			Pogue (PLT)	A7LB-629	12/25, 07:01:01		Carr, Pogue: study Kohoutek;
			Gibson (SPT)	A7LB-628	12/29, 03:28:16		Gibson, Carr: study Kohoutek;
					2/3, 05:28:00		Carr, Gibson: collect experiments

APPENDIX 10
Apollo-Soyuz Test Project Summary

Date of Launch	Date of Landing	Flight	Crew	Suit	Mission Duration hr:min:sec	Objectives/Accomplishments
7/15/75	7/24/75	ASTP	Stafford (CDR) Brand (CMP) Slayton (DMP)	A7LB-801 A7LB-806 A7LB-803	217:28:24 (9 days)	Tested international docking system and joint flight procedures; no EVA

Space Shuttle Summary

Date of Launch	Date of Landing	Landing Site	Flight	Duration hrs:min:sec	Crew	Spacecraft	Payload/Objectives/Accomplishments
4/12/81	4/14/81	Edwards	STS-1	54:20:32 (2 days)	Young (CDR) Crippen (PLT)	Columbia	Test; 36.5 orbits
11/12/81	11/14/81	Edwards	STS-2	54:13:13 (2 days)	Engle (CDR) Truly (PLT)	Columbia	Test; 36 orbits
3/22/82	3/30/82	White Sands	STS-3	192:04:45 (8 days)	Lousma (CDR) Fullerton (PLT)	Columbia	Test; 129 orbits
6/27/82	7/4/82	Edwards	STS-4	169:09:40 (7 days)	Mattingly (CDR) Hartsfield (PLT)	Columbia	Test; 112 orbits; tried on PGA
11/11/82	11/16/82	Edwards	STS-5	122:14:26 (5 days)	Brand (CDR) Overmyer (PLT) Lenoir (MS) Allen (MS)	Columbia	First operational flight; 81 orbits; first launch of 2 satellites; first 4-member crew; EVA canceled, suit problems
4/4/83	4/9/83	Edwards	STS-6	120:23:42 (5 days)	Weitz (CDR) Bobko* (PLT) Musgrave* (MS) Peterson (MS)	Challenger	Maiden voyage; 80 orbits; first EVA, 4/7/83, 4 hr. 17 min.; first TDRS launch
6/18/83	6/24/83	Edwards	STS-7	146:23:59 (6 days)	Crippen (CDR) Hauck (PLT) Fabian (MS) Ride (MS) Thagard (MS)	Challenger	First American female MS (Ride); 97 orbits; 2 commercial satellites; West German satellite released and retrieved by robotic arm
8/30/83	9/5/83	Edwards	STS-8	145:08:43 (6 days)	Truly (CDR) Brandenstein (PLT) Gardner (MS) Bluford (MS) Thornton, W. (MS)	Challenger	First night launch; 97 orbits; first night landing; satellite launched; first African American MS (Bluford)
11/28/83	12/8/83	Edwards	STS-9	247:47:24 (10 days)	Young (CDR) Shaw (PLT) Garriott (MS) Parker (MS) Lichtenberg (MS) Merbold (PS)	Columbia	First 6-person crew; 166 orbits; first spacelab flight; first payload specialist; astronomy, solar physics; first civilian scientist; first European PS (Merbold)
2/3/84	2/11/84	Kennedy (STS-11)	41-B	191:15:55 (7 days)	Brand (CDR) Gibson (PLT) McNair (MS) Stewart* (MS) McCandless* (MS)	Challenger	First KSC landing; 127 orbits; 3 satellites launched; first EVA test of MMU, 2/7; second EVA, 2/9; 12-hr EVA total
4/6/84	4/13/84	Edwards (STS-13)	41-C	167:40:07 (6 days)	Crippen (CDR) Scobee (PLT)	Challenger	Long-duration facility; 107 orbits; EVA: 4/9, almost 6 hr., failed to catch

Launch	Landing	Site	Mission	Duration	Crew	Orbiter	Notes
					Hart (MS) van Hoften* (MS) Nelson* (MS)		Solar Max; EVA: 4/10, 7 hr. 18 min., Solar Max repaired
8/30/84	9/5/84	Edwards	41-D	144:56:04 (6 days)	Hartsfield (CDR) Coats (PLT) Mullane (MS) Hawley (MS) Resnik (MS) Walker, C. (PS)	Discovery	Maiden voyage; 96 orbits; 3 satellites launched; second American female MS (Resnik); first commercial payload specialist; electrophoresis
10/5/84	10/13/84	Kennedy	41-G	197:23:33 (8 days)	Crippen (CDR) McBride (PLT) Sullivan* (MS) Ride (MS) Leestma* (MS) Scully-Power (MS) Garneau (PS)	Challenger	First American female EVA (Sullivan); 132 orbits; first U.S. orbital refueling; first crew of 7; first USN oceanographer (Scully-Power); earth observations; first Canadian MS (Garneau), in return for robotic arm cooperation
11/8/84	11/16/84	Kennedy	51-A	191:45:54 (7 days)	Hauck (CDR) Walker, D. (PLT) Allen* (MS) Fisher, A. (MS) Gardner, D.* (MS)	Discovery	Retrieval of Palapa and Westar; 126 orbits; launched 2 satellites; EVA: 11/13, 4 hr. 30 min.
1/24/85	1/27/85	Kennedy	51-C	73:33:27 (3 days)	Mattingly (CDR) Shriver (PLT) Onizuka (MS) Buchli (MS) Payton (PS)	Discovery	DOD; 47 orbits
4/12/85	4/19/85	Kennedy	51-D	167:55:23 (6 days)	Bobko (CDR) Williams (PLT) Hoffman* (MS) Griggs* (MS) Seddon (MS) Walker, C. (PS) Garn (PS)	Discovery	Launched 2 satellites; 109 orbits; electrophoresis; unplanned EVA; 3 hr., attempt to repair Syncom; first U.S. Senator (Garn)
4/29/85	5/6/85	Edwards	51-B	168:08:46 (7 days)	Overmyer (CDR) Gregory (PLT) Lind (MS) Thagard (MS) Thornton, W. (MS) Wang (MS) van den Berg (PS)	Challenger	Spacelab 3; 110 orbits; second SL mission; life sciences—animal on board; materials science, astrophysics; first Netherlander (van den Berg)
6/17/85	6/24/85	Edwards	51-G	169:38:52 (7 days)	Brandenstein (CDR) Creighton (PLT) Fabian (MS) Nagel (MS)	Discovery	First Saudi Arabian PS (Al-Saud); 112 orbits; first French PS (Baudry); medical tests; photography; launched satellites; laser-beam test

Space Shuttle Summary

Date of Launch	Date of Landing	Landing Site	Flight	Duration hrs:min:sec	Crew	Spacecraft	Payload/Objectives/Accomplishments
7/29/85	8/6/85	Edwards	51-F	190:45:12 (7 days)	Fullerton (CDR) Bridges (PLT) Henize (MS) Musgrave (MS) England (MS) Acton (PS) Bartoe (PS)	Challenger	Spacelab 2; 126 orbits; solar observatory
8/27/85	9/3/85	Edwards	51-I	170:17:42 (7 days)	Engle (CDR) Covey (PLT) van Hoften (MS) Fisher, W.,* (MS) Lounge (MS)	Discovery	Launched satellites; 111 orbits; repaired Syncom IV-4; EVA: 8/31, 7 hr. 8 min.; EVA: 9/1, 4 hr.
10/3/85	10/7/85	Edwards	51-J	97:44:38 (4 days)	Bobko (CDR) Grabe (PLT) Hilmers (MS) Stewart (MS) Pailes (MS)	Atlantis	Maiden voyage; 63 orbits; DOD
10/30/85	11/6/85	Edwards	61-A	168:44:51 (7 days)	Hartsfield (CDR) Nagel (PLT) Dunbar (MS) Buchli (MS) Bluford (MS) Messerschmid (PS) Furrer (PS) Ockles (PS)	Challenger	Spacelab D-1; 110 orbits; first 8-person crew; first German PS's (Messerschmid, Furrer)
11/26/85	12/3/85	Edwards	61-B	165:04:49 (6 days)	Shaw (CDR) O'Connor (PLT) Spring (MS) Cleave (MS) Ross (MS) Walker, C. (PS) Neri Vela (PS)	Atlantis	EVA: 11/29, 6 hr. 30 min.; 108 orbits; EVA: 11/30, 6 hr. 30 min.; EASE/Access construction practice; electrophoresis; launched satellites; first Mexican PS (Neri Vela)
1/12/86	1/18/86	Edwards	61-C	146:03:51 (6 days)	Gibson (CDR) Bolden (PLT) Chang-Diaz (MS) Hawley (MS)	Columbia	Launched satellites; 97 orbits; astrophysics; materials science lab; GAS; first U.S. Representative (B. Nelson)

The main part of the first row (before Challenger/7-29-85) also lists:
Lucid (MS) Baudry (PS) Al-Saud (PS)

Date	Date	Landing	Mission	Duration	Crew	Orbiter	Remarks
1/28/86			51-L	00:01:13	Nelson, G. (MS) Cenker (PS) Nelson, B. (PS) Scobee (CDR) Smith (PLT) Onizuka (MS) Resnik (MS) McNair (MS) McAuliffe (PS) Jarvis (PS)	Challenger	First teacher to fly in space program (McAuliffe); explosion claimed crew and orbiter
9/29/88	10/3/88	Edwards	STS-26	97:00:11 (4 days)	Hauck (CDR) Covey (PLT) Hilmers (MS) Lounge (MS) Nelson, G.	Discovery	Deployed TDRS-C; 63 orbits; 2 student projects; microgravity research; materials processing
12/2/88	12/6/88	Edwards	STS-27	105:06:19 (4 days)	Gibson (CDR) Gardner (PLT) Ross (MS) Mullane (MS) Shepherd (MS)	Atlantis	Deployed Lacrosse DOD satellite; 68 orbits; evaluated carbon-carbon composite material under Shuttle nose
3/13/89	3/18/89	Edwards	STS-29	119:38:52 (4 days)	Coats (CDR) Blaha (PLT) Buchli (MS) Springer (MS) Bagian (MS)	Discovery	TDRS-D/IUS; 79 orbits; life science experiment
5/4/89	5/8/89	Edwards	STS-30	96:57:33 (4 days)	Walker, D. (CDR) Grabe (PLT) Thagard (MS) Cleave (MS) Lee (MS)	Atlantis	64 orbits; Magellan spacecraft to Venus
8/8/89	8/13/89	Edwards	STS-28	121:00:09 (5 days)	Shaw (CDR) Richards (PLT) Leestma (MS) Brown (MS) Adamson (MS)	Columbia	DOD; 81 orbits; deployed and tested advanced imaging satellite
10/18/89	10/23/89	Edwards	STS-34	119:39:24 (4 days)	Williams (CDR) McCulley (PLT) Lucid (MS) Baker, E. (MS) Chang-Diaz (MS)	Atlantis	Galileo; 79 orbits
11/22/89	11/27/89	Edwards	STS-33	120:06:49 (5 days)	Gregory (CDR) Blaha (PLT) Thornton, K. (MS) Carter (MS) Musgrave (MS)	Discovery	DOD; 78 orbits

Space Shuttle Summary

Date of Launch	Date of Landing	Landing Site	Flight	Duration hrs:min:sec	Crew	Spacecraft	Payload/Objectives/Accomplishments
1/9/90	1/20/90	Edwards	STS-32	261:00:37 (10 days)	Brandenstein (CDR) Wetherbee (PLT) Dunbar (MS) Low (MS) Ivins (MS)	Columbia	Deployed Syncom; 171 orbits; retrieve LDEF
2/28/90	3/4/90	Edwards	STS-36	106:18:23 (4 days)	Creighton (CDR) Casper (PLT) Thuot (MS) Mullane (MS) Hilmers (MS)	Atlantis	DOD; 72 orbits
4/24/90	4/29/90	Edwards	STS-31	121:16:05 (5 days)	Shriver (CDR) Bolden (PLT) Hawley (MS) McCandless (MS) Sullivan (MS)	Discovery	76 orbits; deployed Hubble telescope
10/6/90	10/10/90	Edwards	STS-41	98:10:00 (4 days)	Richards (CDR) Cabana (PLT) Melnick (MS) Shepherd (MS) Akers (MS)	Discovery	Ulysses/IUS/PAM-S; 65 orbits; Intelsat Solar Ar. Coupon; SSBUV; CHROMEX
11/15/90	11/20/90	Kennedy	STS-38	117:55:22 (4 days)	Covey (CDR) Culbertson (PLT) Gemar (MS) Meade (MS) Springer (MS)	Atlantis	DOD; 79 orbits
12/2/90	12/10/90	Edwards	STS-35	215:05:08 (8 days)	Brand (CDR) Gardner, G. (PLT) Hoffman (MS) Lounge (MS) Parker (MS) Durrance (PS) Parise (PS)	Columbia	HUT; WUPPE; UIT; 142 orbits; BBXRT; Astro-One
4/5/91	4/11/91	Edwards	STS-37	143:32:44 (5 days)	Nagel (CDR) Cameron (PLT) Godwin (MS) Ross* (MS) Apt* (MS)	Atlantis	GRO; 93 orbits; EVA: 4/7, 4 hr. 30 min.; EVA: 4/8, 6 hrs., 11 min.

Date	Landing	Mission	Duration	Crew	Orbiter	Payload
4/28/91	Edwards	STS-39	199:22:22 (8 days)	Coats (CDR) Hammond (PLT) Harbaugh (MS) McMonagle (MS) Bluford (MS) Veach (MS) Hieb (MS)	Discovery	134 orbits; IBSS, SPAS
6/5/91	Edwards	STS-40	218:14:20 (9 days)	O'Connor (CDR) Gutierrez (PLT) Bagian (MS) Jernigan, (MS) Seddon (MS) Gaffney (PS) Hughes-Fulford (PS)	Columbia	SLS-1; 146 orbits; 12 GAS; 7 OEX
8/2/91	Kennedy	STS-43	213:21:25 (8 days)	Blaha (CDR) Baker, M. (PLT) Lucid (MS) Low (MS) Adamson (MS)	Atlantis	TDRS-E/IUS; 142 orbits; SHARE-II; SSBUV; polymer membrane processing
9/12/91	Edwards	STS-48	128:27:00 (5 days)	Creighton (CDR) Reightler (PLT) Gemar (MS) Buchli (MS) Brown (MS)	Discovery	UARS; 81 orbits; APM-3; protein crystal growth
11/24/91	Edwards	STS-44	166:50:00 (6 days)	Gregory (CDR) Henricks (PLT) Voss (MS) Musgrave (MS) Runco (MS) Hennen (PS)	Atlantis	DSP/IUS; 109 orbits; IOCM; Terra Scout; AMOS; CREAM; RME-III
1/22/92	Edwards	STS-42	193:14:00 (8 days)	Grabe (CDR) Oswald (PLT) Thagard (MS) Readdy (MS) Hilmers (MS) Bondar (PS) Merbold (PS)	Discovery	IML-1; 129 orbits; GOSAMR-1; IPMP; RME-III; SE-81-09; GAS Bridge; first Canadian female PS (Bondar)
3/24/92	Kennedy	STS-45	214:09:00 (8 days)	Bolden (CDR) Duffy (PLT) Sullivan (MS) Leestma (MS)	Atlantis	ATLAS-1; 143 orbits; SSBUV; RME-II; VFT-II; CLOUDS-1A; SAREX; IPMP

Space Shuttle Summary

Date of Launch	Date of Landing	Flight	Duration hrs:min:sec	Landing Site	Crew	Spacecraft	Payload/Objectives/Accomplishments
5/7/92	5/16/92	STS-49	213:17:00 (8 days)	Edwards	Brandenstein (CDR) Chilton (PLT) Thuot* (MS) 3 EVAs Thornton, K.* (MS) 1 EVA Hieb* (MS) 3 EVAs Akers* (MS) 2 EVAs Melnick (MS)	Endeavour	First time four EVAs completed: 3 hr. 43 min.; 5 hr. 30 min; 8 hr. 29 min. (longest EVA); 7 hr. 45 min.; first triple EVA: 5/14/92; maiden voyage; 141 orbits; Intelsat VI rendezvous, repair, reboost; second American female EVA (Thornton); space station assembly practice
6/25/92	7/9/92	STS-50	331:30:00 (13 days)	Kennedy	Richards (CDR) Bowersox (PLT) Dunbar (MS) Baker, E. (MS) Meade (MS) DeLucas (PS) Trinh (PS)	Columbia	221 orbits; U.S. microgravity laboratory; crystal growth fluid physics experiments; astroculture; combustion science; biotechnology experiments
7/31/92	8/8/92	STS-46	191:16:00 (7 days)	Kennedy	Shriver (CDR) Allen (PLT) Hoffman (MS) Chang-Diaz (MS) Nicollier (MS) Ivins (MS) Malerba (PS)	Atlantis	EURECA; 127 orbits; Tether Satellite System; EURECA science; solution growth protein crystallization; first Swiss MS (Nicollier); first Italian PS (Malerba)
9/12/92	9/20/92	STS-47	190:23:00 (7 days)	Kennedy	Gibson (CDR) Brown (PLT) Lee (MS) Apt (MS) Davis (MS) Jemison (MS) Mohri (PS)	Endeavour	Spacelab-J; 127 orbits; materials science; life science; NASDA science; Israeli hornets investigation; amateur radio; solid surface combustion experiment; first Japanese PS (Mohri)
12/2/92	12/9/92	STS-53	173:55:00 (7 days)	Edwards	Walker, D. (CDR) Cabana (PLT) Bluford (MS) Voss (MS) Clifford (MS)	Discovery	DOD; secondary payloads; cryogenic heat pipe experiments; GAS experiments; BLAST

Foale (MS)
Frimout (PS)
Lichtenberg (PS)

*EVA astronaut

Note: Specific suit designations are not listed because Shuttle suits are reusable; therefore no astronaut has any assigned suit.

In 1984, NASA adopted a new numbering system for payload assignments. The first digit represented the fiscal year (4), the second digit designated the launch site (1 = Kennedy Space Center [KSC], 2 = Vandenberg Air Force Base [VAFB]). The sequence of missions in a year was designated by the letters. In 1988, NASA resumed the original method of numbering payload assignments. Sequence of Shuttle flights often changed due to delays or postponements.

Space Shuttle Missions

Columbia: 1, 2, 3, 4, 5, 9, 61C, STS-28, STS-32, STS-35, STS-40, STS-50
12 missions

Challenger: 6, 7, 8, 41-B, 41-C, 41-G, 51-B, 51-F, 61-A, 51-L
10 missions

Discovery: 41-D, 51-A, 51-C, 51-D, 51-G, 51-I, STS-26, STS-29, STS-33, STS-31, STS-41, STS-39, STS-48, STS-42, STS-53
15 missions

Atlantis: 51-J, 61-B, STS-27, STS-30, STS-34, STS-36, STS-37, STS-43, STS-44, STS-45, STS-46
12 missions

Endeavour: STS-49, STS-47
2 missions

In October of 1992, the astronaut corps totaled 89. At the time of the *Challenger* explosion, there were eight women, four African American males, eight non-U.S. citizens, two Congressional observers, and two USAF specialists in the astronaut corps. Two women, Judy Resnik and Christa McAuliffe, died aboard *Challenger* Mission 51-L. Sally Ride resigned from the astronaut corps in the fall of 1987. Mary Cleave and Kathryn Sullivan have also left NASA for other assignments. Fifteen more women have since joined the astronaut ranks: Jan Davis, Ph.D; Marsha Ivins; Mae Jemison, M.D. (first African American female); Tammy Jernigan, Ph.D.; Ellen Shulman-Baker, M.D.; Kathryn Thornton, Ph.D.; Linda Godwin, Ph.D.; Air Force Major Eileen Collins (first female pilot); Army Captain Nancy Sherlock; Air Force Major Susan Helms; Ellen Ochoa, Ph.D. (first Hispanic female); Janice Voss, Ph.D.; U.S. Air Force Captain Catherine G. Coleman; U.S. Navy Lieutenant Commander Wendy B. Lawrence; and Mary E. Weber, Ph.D. Of the 23 women, 15 have flown at least once on the Space Shuttle. There have also been three female payload specialists: Millie Hughes-Fulford, Roberta Bondar, from Canada, and Christa McAuliffe.

Much of the information contained in these charts was provided by Ed Campion of the NASA Public Affairs Office.

Abbreviations

AES	advanced extravehicular suit
AFSC	Air Force Systems Command
ALSA	astronaut life support assembly
ALSEP	Apollo Lunar Surface Experiments Package
AMOS	Air Force Maui Optical System
AMU	astronaut maneuvering unit
APM	atmospheric particle monitor
ARDC	Air Research and Development Command
ASTP	Apollo-Soyuz Test Project
ATDA	augmented target docking adapter
ATLAS-1	Atmospheric Laboratory for Applications and Science-1
ATM	Apollo telescope mount
BBXRT	Broad-Band X-Ray Telescope (payload)
BLAST	Battlefield Laser Acquisition Sensor Test
BSLSS	buddy secondary life support system/backup
BUAER	Bureau of Aeronautics
CAL	Conservation Analytical Lab
CCA	communications carrier assembly
CDR	commander
CFUAS	contingency female urine absorption system
CHROMEX	Chromosome and Plant Cell Division in Space
CLOUDS-1A	Cloud Logic to Optimize Use of Defense Systems
CM	command module
CMD PLT	command pilot

CMP	command module pilot
CP	command pilot
CREAM	Cosmic Radiation Effects and Activation Monitor
CSM	command and service module
CWG	constant wear garment
DACT	disposable absorbent containment trunk
DCM	display and control module
DMP	docking module pilot
DOD	Department of Defense
DSP/IUS	Defense Support Program/inertial upper stage
EASE	Experimental Assembly of Structures in EVA
ECS	environmental control system
EIS	emergency intravehicular suit
ELSA	environmental life support assembly
ELSS	extravehicular life support system
EMU	extravehicular mobility unit
EOS	emergency oxygen system
ESR	Electron Spin Resonance
EURECA	European Carrier Satellite
EV	extravehicular (outside the spacecraft)
EVA	extravehicular activity (walk in space)
EVVA	extravehicular visor assembly
FCS	fecal containment system
GAS	Get Away Special (experiments)
GOSAMR	Gelation of Sols: Applied Microgravity Research
GRO	Gamma Ray Observatory
GT	Gemini Titan
HMA	high mobility arm
HS	Hamilton Standard
HT	high temperature
HUT	hard upper torso, Hopkins Ultraviolet Telescope (payload)
IBSS	Infrared Background Signature Survey
ICBM	intercontinental ballistic missile
ICG	inflight coverall garment
ICL	intravehicular cover layer
IDB	in-suit drink bag
IDS	international docking system
ILC	International Latex Cooperation
IML	International Microgravity Lab
IOCM	Interim Operational Contamination Monitor
IPMP	Investigations into Polymer Membrane Processing
ITLSA	integrated torso-limb suit assembly
ITMG	integrated thermal micrometeoroid garment
IV	intravehicular (inside the spacecraft)

IVA	intravehicular activity
JSC	Johnson Space Center
LAAP	lunar advanced Apollo program
LCG	liquid cooling garment
LCVG	liquid cooling and ventilation garment
LDEF	Long Duration Exposure Facility
LES	launch/entry suit
LEVA	lunar extravehicular activity
LEVVA	lunar extravehicular visor assembly
LM	lunar module
LMP	lunar module pilot
LR	lunar rover
LSS	life support system
LTA	lower torso assembly
MA	Mercury Atlas
MET	modularized equipment transporter
MMU	manned maneuvering unit
MODE	Mid-Deck Zero-Gravity Dynamics Experiment
MOL	Manned Orbiting Laboratory
MR	Mercury Redstone
MS	mission specialist
MSC	manned spacecraft center
MWS	mini-work station
NACA	National Advisory Committee for Aeronautics
NACEL	Naval Air Crew Equipment Laboratory
NASA	National Aeronautics and Space Administration
NASDA	National Space/Development Agency of Japan
NASM	National Air and Space Museum
OES	orbital extravehicular suit
OEP	Orbiter Experiments Program (flight research experiments)
OPS	oxygen purge system
P	pilot
PARE	Physiological and Anatomical Rodent Experiment
PCG	Protein Crystal Growth
PFR	portable foot restraint
PGA	pressure garment assembly
PLSS	portable life support system, primary life support system
PLT	pilot
PS	payload specialist
RME	Radiation Monitoring Experiment
RMS	remote manipulator system
SAREX	Shuttle Amateur Radio Experiment
S/C	spacecraft
SE-81-09	Student Experiment, Convection in Zero Gravity

SE-82-03	Student Experiment, Capillary Rise of Liquid through Granular Porous Media
SEVA	standup EVA
SEVVA	Skylab extravehicular visor assembly
SHARE	Space Station Heat Pipe Advanced Radiator Element
SLS	Spacelab Life Sciences (studies regarding adaptation of humans to space)
SOP	secondary oxygen pack
SPAS	Shuttle Pallet Satellite I
SPT	science pilot
SSA	space suit assembly
SSBUV	Shuttle Solar Backscatter Ultraviolet Instrument (payload)
STL	space tissue loss
STS	Space Transportation System
TDRS-E/IUS	Tracking and Data Relay Satellite-E/inertial upper stage
TERA Scout	Earth Observation Experiment
TLSA	torso-limb suit assembly
TMG	thermal micrometeoroid garment
UARS	Upper Atmospheric Research Satellite
UCD	urine collection device
UCTA	urine collection transfer assembly
UIT	Ultraviolet Imaging Telescope (payload)
VFT	visual function tester
WETF	Weightless Environment Training Facility
WUPPE	Wisconsin UV Photo-Polarimeter Experiment (payload)
ZPS	zero-prebreathe suit

Notes

Chapter 1. Flying Suits

1. Tom Crouch, *The Bishop's Boys,* pp. 187–286. Tom Crouch, *The Eagle Aloft,* p. 593. Orville Wright, "How We Made the First Flight," pp. 105–9; "Amos I. Root Sees Wilbur Wright Fly," pp. 110–15 from *The Wright Brothers Heirs of Prometheus,* Richard P. Hallion, ed. (Washington, D.C.: Smithsonian Institution Press, 1978). Lloyd S. Swenson, Jr., James M. Grimwood, and Charles C. Alexander, *This New Ocean,* pp. 5–13. E. W. Still, *Into Thin Air,* p. 5. Also see Homer E. Newell, *Beyond the Atmosphere,* p. 63. I am very grateful to G. Harry Stine for calling to my attention his book *The Handbook for Space Colonists,* illustrated by Rick Sternbach, pp. 50, 52, 57, and 59. The information is clearly written and easily understood. Figures 1. 2 and 1.3 have been adapted from his book.

2. Eloise K. Engle and Arnold S. Lott, *Man in Flight,* pp. 39–77. Louis Lasagna, "With a Wing and a Prayer," *The Sciences* (Mar.–Apr. 1985):8.

3. Martin and Grace Caidin, *Aviation and Space Medicine,* pp. 58–69, 90, and 123. Also see John W. R. Taylor, *Air Facts and Feats,* p. 231.

4. John Scott Haldane and J. G. Priestly, *Respiration,* p. 322. Also see Engle and Lott, *Man in Flight,* pp. 38–40, and Marvin Miles, "Men 'Soar' 89 Miles—on Ground," *Los Angeles Times,* Oct. 7, 1957, part 1, p. 1.

5. C. G. Sweeting, *Combat Flying Clothing,* pp. 72–84. Also see Engle and Lott, *Man in Flight,* pp. 218–19, and Adrianne Noë, pers. corr. to the author, May 12, 1987.

6. John D. Anderson, Jr., *Introduction to Flight* (New York: McGraw-Hill, Inc., 1978), p. 340.

7. John A. Macready, "Airplane Reaches Altitude of 7.75 miles," in *Test Pilots,* pp. 17–21.

8. Lloyd Mallan, *Suiting Up for Space,* pp. 1–13.

9. Wiley Post, "Wiley Post," *Popular Mechanics* (Oct. 1934):492–95.

10. T. W. Walker, "The Development of the Pressure Suit for High Altitude Flying," *The Project Engineer,* 15 (May 1956):4. Also see Stanley R.

Mohler and Bobby H. Johnson, *Wiley Post, His Winnie Mae, and the World's First Pressure Suit,* pp. 71–101.

11. Charles Wilson, "Wiley Post," unedited article for *Archives of Environmental Health* (in NASM Archives).

12. Haldane and Priestley, *Respiration,* pp. 325–26.

13. Randy Smith "Post and Rogers—Se-Tenants in Life," *Stamp World* (Aug. 1983): 62. Mohler and Johnson, *Wiley Post,* pp. 81–102, Mallan, *Suiting Up for Space,* p. 33.

14. Russell S. Colley, oral history interview with the author and George Kydd, of the Naval Air Development Center, Warminster, Penn., at Pinehurst, N.C., May 9, 1989 (transcript in NASM Archives). Robert A. Brown, pers. corr. to Smithsonian, May 9, 1985 (in NASM Archives). Helen W. Schulz, "Case History of Pressure Suits," p. 19 (Digest). Also see J. A. Roth, "Report on Pressure Altitude Suit Equipment," *AAF Proving Ground Command Report,* p. 5. C. G. Sweeting, pers. comm., May 20, 1987. Swenson, et al., *This New Ocean,* pp. 3–15. Also see Sweeting, *Combat Flying Equipment,* pp. 166–80.

15. Wayne K. Galloway, pers. corr. to the author, September 8, 1985 (in NASM Archives). Also see Sweeting, *Combat Flying Equipment,* pp. 164–77.

16. Theodore von Kármán with Lee Edson, *The Wind and Beyond: Theodore von Kármán* (Canada: Little, Brown & Company, 1967) pp. 218–21, *Naval Academy News* (Mar. 1959):30. Also see Engle and Lott, *Man in Flight,* p. 77.

17. David Baker, *The History of Manned Space Flight,* pp. 12 and 13. James Gunn, *Alternate Worlds, The Illustrated History of Science Fiction* (New Jersey: Prentice-Hall, Inc., 1975), p. 76. Sam Moskowitz, *The Immortal Storm* (Atlanta, Ga.: The Atlanta Science Fiction Organization Press, 1954), p. 17. *Toy World* (Oct. 1934), 8:47.

Chapter 2 Development of the Pressure Suit

1. Walter A. McDougall, . . . *the Heavens and the Earth* (New York: Basic Books, Inc., 1985), pp. 48–143. Peter Young, ed.-in-chief, *Illustrated World War II Encyclopedia* (Westport, Conn.: H. S. Stuttman Inc., 1978), pp. 1040–49. Engle and Lott, *Man in Flight,* p. 227.

2. Alice King Chatham, "Jacket and Swivel Tethering Systems," reprint from *Lab Animal,* Nov.–Dec., 1985. H. J. Feldman, "Space Age Sculptress Alice K. Chatham," *Satellite,* Santa Monica Chapter, McDonnell Douglas, Apr., 1971, vol. 5, no. 4, p. 1. Helen W. Schulz, "Case History of Pressure Suits," pp. 2–3. Capstans were first described by H. Lamport, L. P. Herrington, and E. C. Hoff in "Review of Methods of Applying Air Pressure to the Extremities for Protection Against Acceleration with Measurement of the Effective Pressure on the Skin," Air Materiel Command Report #228, Office of Scientific Research and Development, November 24, 1943.

3. David M. Clark, "On Partial Pressure Suits," from "Space Suit History" compiled by C. C. Lutz, Crew Systems Div., NASA/Houston, pp. 3–12. Also see *Air Force Physiological Training Program News Letter,* No. 44, June 1961, AFCSG-11, in the same compilation (in NASM Archives).

4. J. Allen Neal, "High Altitude Pressure Suits" in *History of Wright Air Development Center,* pp. 8–10 (in NASM Archives).

5. Ibid., p. 20. David M. Clark, "The Development of the Partial Pressure Suit and Something of its Value (as Remembered by Myself, Dave Clark): Part 1," *Aviation, Space, and Environmental Medicine* (June 1989):624–26. (I am very grateful to Dr. George Kydd, formerly of the Naval Air Development Center in Warminster, Penn., for bringing this article to my attention.) Clark, "On Partial Pressure Suits," pp. 5–10.

6. Lloyd Mallan, *Suiting Up for Space,* pp. 122–23. Also see *Air Force Physiological Training Program News Letter,* No. 44, June 1961, p. 5.

7. Neal, "High Altitude Pressure Suits," pp. 27–31.

8. A. Scott Crossfield, *Onward and Upward* (Bismarck, N. Dak.: Maurice E. Cook, 1988), p. 12. "USAF Full Pressure Space Suit is Light, Permits Free Movement," n.a., *Aviation Week* (Dec. 9, 1957):29.

9. U.S. Dept. of the Navy, Bureau of Aeronautics, "Buaer Aviation Clothing and Survival Equipment Bulletin No. 1–59." AER-AE-5111/25, June 18, 1959, p. 3.

10. Mallan, *Suiting Up for Space,* pp. 134–49.

11. Charles F. Gell, Edward L. Hays, and James V. Correale, "Developmental History of the Aviator's Full Pressure Suit in the U.S. Navy," *The Journal of Aviation Medicine,* 30 (Apr. 1959):242–48.

12. Mallan, *Suiting Up for Space,* p. 140.

13. *Naval Aviation News* (1958):2, and Gell et

al., "Developmental History . . . ," pp. 246–48.
Also see pers. comm. from Dennis Gilliam, July 29,
1986, and Wayne Galloway, August 5, 1986 (in
NASM Archives).

14. *Naval Academy News* (March 1959):30.
Also see Navy Training Courses, *Parachute Rigger 1
& C,* Bureau of Naval Personnel, NAVPERS 10360,
1959, pp. 373–75.

Chapter 3 Mercury Space Suits

1. "National Aeronautics and Space Act of
1958," Public Law 85-568, 85th Cong., HR 12575,
July 29, 1958.

2. Lloyd S. Swenson, Jr., James M. Grimwood,
and Charles C. Alexander, *This New Ocean,* pp.
14–31 and 68–78. Ruth G. von Sauma and Walter
Wiesman, "The German Rocket Team," Alabama
Space and Rocket Center, August 1983, pp. 5 and 6.
K. Greengouse, "Science or Ambitions?" *Science in
the USSR* (May–June 1990):1.128, no. 3, p. 21.

3. John Dille, ed., *We Seven,* pp. 110–16. Engle
and Lott, *Man in Flight,* pp. 81–105.

4. Frank H. Winter, *Prelude to the Space Age,
The Rocket Societies: 1924–1940* (Washington,
D.C.: Smithsonian Institution Press, 1983), p. 25.

5. Russell Colley, Barbara Fludine, Department
of Space History Oral History Project, May 9 and
10, 1989 (in NASM Archives). Swenson et al., *This
New Ocean,* p. 228. Also see L. N. McMillion, Life
Systems Branch, memo to Chief, Flight Systems Di-
vision, December 31, 1959. Also see Engle and
Lott, *Man in Flight,* p. 231. Also see "Each $5000
Pressure Suit Requires a Month to Make," special to
Buffalo Evening News, approximate date of July
1961 handwritten on the clipping (in NASM
Archives).

6. L. N. McMillion, "Pressure Suit Status Report," memo to Chief, Flight Systems Division, December 24, 1959. Also McMillion, "Trip Report,"
memo to Chief, Flight Systems Division, December
31, 1959.

7. Walter Schirra, "A Suit Tailor-Made for
Space," *Life* (Aug. 1, 1960): 36.

8. Members of the NASA Space Task Group usually included the seven astronauts: Lee N.
McMillion from NASA, Life Systems Branch; James
Correale and Lester Snider, space suit designers
from the Navy Bureau of Astronautics; and various
B. F. Goodrich people such as Raymond Anderson,
Richard Shaw, Paul Frecka, Wayne Galloway, Donald Wohlgemuth, Carl Effler, Donald Ewing, and
others. The most senior Goodrich representative
and head designer was still Russell Colley, the cre-
ator of Wiley Post's first successful pressure suit.

9. L. N. McMillion, Life Systems Branch, "Astronaut Comments on Pressure Suit," memo to Max-
ime A. Faget, Chief, Flight Systems Division, June
27, 1960. Also see Swenson et al., *This New Ocean,*
p. 230.

10. Dille, *We Seven,* p. 154.

11. Swenson et al., *This New Ocean,* p. 484.

12. Stanley C. White, "Vomiting in Helmet,"
memo to Life Systems Branch, December, 1961.

13. Swenson et al., *This New Ocean,* p. 500.

14. Ibid., pp. 509–11.

Chapter 4 Gemini Space Suits

1. Walter A. McDougall, . . . *the Heavens and
the Earth* (New York: Basic Books, Inc., 1985), pp.
332–402. James V. Correale and Walter W. Guy,
"Space Suits," *Space World,* Vol. A-1 (Sept.–Oct.
1963):16. Also see Richard S. Johnston, Correale,
and Matthew I. Radnofsky, *Space Suit Development,* NASA TN D-3291 (Washington, D.C.,
1966), pp. 1–2; and James M. Grimwood, Barton
C. Hacker, with Peter J. Vorzimmer, *Project Gemini,
A Chronology;* pp. 12–19, 37–38, 60, and 283.
Also see pers. corr. from Joseph Ruseckas to author,
January 20, 1986 (in NASM Archives).

2. Project Gemini, Extravehicular Pressurization
Ventilation System, Statement of Work, MSC, June
19, 1963. Ivan D. Ertel, "Gemini Program," NASA
Fact Sheet 291-F, August, 1966. Also see NASA,
Manned Spacecraft Center, News Release MSC
63-110, July 5, 1963. Warren Burkett, "Eight Gem-
ini Astronaut Suits Begun," *Missiles and Rockets*
(May 25, 1964):31; Lloyd Mallan, *Suiting Up for
Space,* pp. 142–194; and NASA, "Statement of
Work, Gemini Space Suit Reliability Testing Pro-
gram," December 2, 1963.

3. Grimwood et al., *Project Gemini Chronology,*
pp. 63, 84, 103, 126, 127. David Clark Company,
"Technical Manual Maintenance, Repair and Test-
ing of Space Suit Type G-2C"; and NASA MSC
"Field Maintenance Instructions for Gemini Space
Suit Type G-3C," Apr.–Oct., 1964. para. 2.1.6;
Jonathan Eberhart, "U.S. Astronauts Could Fly
without Space Suits," Science Service News Release,
October 13, 1964.

4. J. T. Barker, "Project Gemini Extra-Vehicular

Pressurization Ventilation System Statement of Work," Crew Equipment Branch, MSC, June 19, 1963, pp. 1–2. Also see NASA, MSC, "Design Certification Review, GT-3, Gemini Intravehicular Space Suit Assembly," February 1965. David Clark Company, "Report of New Technology, Project Gemini Space Suit Program," May 11, 1964. David Clark Company, "Field Maintenance Instructions for Gemini Space Suit Type G-3C," October 1964, pp. 10–13.

5. David Clark Company, "Gemini EVA Space Suit Assembly Familiarization Handbook," N.D. pp. 4–26.

6. R. M. Machell et al., "Crew Station and Extravehicular Equipment," *Gemini Midprogram Conference*, p. 69. Also see NASA, MSC, "Gemini 4 Flight," Fact Sheet 291-B, 1965. David Clark Company "Statement of Work for Hazardous Environment Protective Assembly (S-1C)" (in NASM Archives). Also, "Meteoroid Qualification of the Gemini Space Suit, GT-8," memo from NASA Meteoroid Technology and Optics Branch to NASA Gemini Support Office, February 4, 1966.

7. NASA, Manned Spacecraft Center, "Gemini VII/Gemini VI," Fact Sheet 291-D, January 1966.

8. E. M. Tucker, "GT-6 Post-Flight Report, Space Suits and Accessories," January 4, 1966.

9. MSC, "Design Certification Report on G-5C Space Suit for GT-7 Mission," n.d. Also see Tucker, "GT-7 Space Suit Post Flight Report," January 4, 1966.

10. NASA, MSC, "Gemini VIII Rendezvous and Docking Mission," Fact Sheet 291-E, April 1966.

11. Larry E. Bell, "Space Suit Configuration for Gemini Missions," memo to Tim White, Spacecraft Integration Branch, MSC, July 27, 1966.

12. Richard S. Johnston and Edward L. Hays, "The Development and Operation of Extravehicular Equipment," 1966. NASA, MSC, "Gemini IX-A Rendezvous Mission," Fact Sheet 291-F, August 1966.

13. Ibid.

14. NASA, MSC, "Gemini X, Multiple Rendezvous, EVA Mission," Fact Sheet 291-G, September 1966.

15. Tucker, "Gemini X Postflight Space Suit Evaluation," Gemini Space Suit Office, August 2, 1966. NASA, MSC, "Gemini X Flight Suit History and Configuration Flight Readiness Review," July 5, 1966.

16. Johnson and Hays, "The Development and Operation of Extravehicular Equipment," pp. 12–

14. NASA, MSC, Flight Readiness Review, "Gemini IX Flight Suit History and Configuration," May 2, 1966.

17. NASA, MSC, "Gemini XI Mission, High Altitude, Tethered Flight," Fact Sheet 291-H, October 1966.

18. Dennis Gilliam pers. comm. to author, December 12, 1986. Also see David Baker, *The History of Manned Spaceflight*, p. 246. F. R. Spross, "Gemini XII Postflight Space Suit Evaluation," Gemini Space Suit Office, November 23, 1966.

19. D. W. Morris, Jr., "Heat Leak Analysis for Gemini Space Suit Utilizing Quilted Superinsulation," memo to Alton Tucker, Gemini Support Office, July 18, 1966.

20. Johnston and Hays, "The Development and Operation of Extravehicular Equipment," 1966.

Chapter 5   Apollo Space Suits

1. Courtney G. Brooks, James M. Grimwood, Loyd S. Swenson, Jr., *Chariots for Apollo*, pp. 4–268. Also see Walter A. McDougall, . . . *the Heavens and the Earth* (New York: Basic Books, Inc., 1985), pp. 381–422; and Edward C. Ezell, "The Apollo Program: History Must Judge," in *Apollo Ten Years Since Tranquillity Base*, edited by Richard P. Hallion and Tom D. Crouch (Washington, D.C.: Smithsonian Institution Press, 1979), pp. 27–34.

2. Wernher von Braun, "What an Astronaut Will Wear on the Moon," *Popular Science*, April, 1965, pp. 87, 88. Eloise K. Engle, and Arnold S. Lott, *Man in Flight*, p. 210, 247.

3. "Apollo Space Suit Contract Issued," *Aviation Week & Space Technology*, October 22, 1962, p. 33.

4. Interview with Matthew I. Radnofsky and James M. Grimwood on the development of the Apollo space suit at Houston Manned Spacecraft Center, March 21, 1966, unpublished transcript, pp. 2, 3.

5. Brooks et al., *Chariots for Apollo*, p. 178, 179. Also see Radnofsky/Grimwood interview, pp. 4, 5.

6. Brooks et al., *Chariots for Apollo*, p. 217.

7. Ibid., p. 221.

8. Radnofsky/Grimwood interview, p. 7. Also see United Press International, *Gemini*, p.16; and "NASA to Negotiate for Apollo Suit, Support System," News Release 65-346; and interview with

Richard Wilde, Extravehicular Systems Engineering Manager, United Technologies Hamilton Standard, December 12, 1986, and Brooks et al., *Chariots for Apollo,* p. 296.

9. I am indebted to Kathy Dirks of the Museum of American History Textile Division, Washington, D.C., for bringing the historical note about "Style 3715" socks (Apollo lunar socks) to my attention. The note concerns a newly accessioned group of socks donated to the Smithsonian Institution by the May & Athens Hosiery Mills, a division of Wayne-Gossard Corporation. A. L. Marcum and H. A. Mauch, "Experimental Space Worker's Garment and Helmet Assembly," pp. 4–13.

10. International Latex Corporation, "Material Specifications for Nylon Spandex, 5 and 6 ounces." Also "Material Specification for Nylon Tricot" and "Material Specification for Tubing, Plastic, Flexible, Modified PVC."

11. Trade names were Lexan, Merlan, or Homolite (Radnofsky/Grimwood interview). Snoopy caps were named for their resemblance to the flight helmet worn by Snoopy in the comic strip "Peanuts." "Astronauts Show Off New Fireproof Space Suit," *The New York Times,* August 3, 1978 (in NASM Archives). Also see Lutz, "Space Suits," p. 13. Radnofsky/Grimwood interview, March 21, 1966, p. 9. Also see Warren G. Wetmore, *Aviation Week & Space Technology,* March 3, 1969, p. 57. And "Apollo Extravehicular Mobility Unit Design and Performance Specification," NASA Document MSC-CSD-A-017, p. 3–102.

12. Wetmore, "Improved Suit Proposed," *Aviation Week & Space Technology,* March 3, 1969, p. 57. Also NASA Fact Sheet, 91-F, August 1966.

13. ILC Industries, Inc., "Space Suits for Project Apollo," 1969. Also see Richard S. Johnston, James V. Correale, and Matthew I. Radnofsky, "Space Suit Development Status," NASA TN D-3291, p. 20. E. I. du Pont de Nemours & Co., "Summary of Properties, Dupont Mylar Polyester," Technical Information, Bulletin M-1C; "Summary of Properties, Dupont H Film polyimide film," Technical Information, Bulletin H-1; and "Dupont Kapton Polyimide Film," May 1, 1966.

14. Wetmore, "Improved Suit Proposed," *Aviation Week & Space Technology,* p. 55. Also International Latex Corporation, "Material Specifications for Cloth, Nomex, 6 oz.," August 25, 1966, p. 2.

15. ILC, "Material Specification for Cloth, Ripstop Nylon, Neoprene Coated, 3.5 oz.," March 29, 1968, ILC, Dover, Delaware. Also ILC, "Material

Specification for Cloth, Polyester, Non-woven," July 13, 1971, ILC, Dover, Delaware.

16. Hamilton Standard, "News," December 6, 1965, p. 3. Stanley M. Luczkowski, "Bioinstrumentation," in *Biomedical Results of Apollo,* Richard S. Johnson, Lawrence F. Dietlein, M.D., and Charles A. Berry, M.D., editors. NASA-SP-368, (Washington, D.C., 1975), p. 496.

17. Hamilton Standard, "Performance and Design Requirements for the Extra-Vehicular Mobility Unit," Apollo Program General Specifications, pp. 3.0–3.1.2. Also see Warren G. Wetmore, "Improved Suit Proposed for Lunar Wear," *Aviation Week & Space Technology,* March 3, 1969. p. 58. Also see "Water-Cooled Undies," *Mechanix Illustrated,* December, 1964, p. 85. Also "Water-Cooled Apollo Suit Prototype Shipped to NASA," *Aviation Week,* March 23, 1964, n.p. (in NASM Archives). And C. C. Lutz, "Space Suits and Support Equipment," February 7, 1966, p. 13.

18. Maurice A. Carson, Michael N. Rouen, Charles C. Lutz, James W. McBarron, II, "Extravehicular Mobility Unit," in *Biomedical Results of Apollo,* edited by Johnson et al., pp. 546, 553. Also see William T. Keovsk, "ILC's customers include the men on the moon," *Prospect '71, The Sunday Bulletin* (Philadelphia, Penn.) October 18, 1970, p. 43. And Fabric Research Laboratories, Inc., "NASA Contract to FRL," NASA News Release, May 3, 1965. "Pressure Garment Assemblies required for NAA," S&ID Apollo Test Program, Enclosure I, NASA, n.d., n.a. (in NASM Archives). Also see Heather M. David, "Apollo Suit Substantially Redesigned," *Missiles and Rockets,* April 26, 1965, pp. 26, 27.

19. "Apollo 9," NASA Mission Report, MR-3, March 26, 1969. "Apollo 9 Verifies Lunar Suit, Backpack," *Aviation Week & Space Technology,* March 17, 1969, p. 67.

20. Michael Collins, *Carrying the Fire,* pp. 359–403. Brooks et al., *Chariots for Apollo,* pp. 361–66.

21. "Apollo-12," NASA Mission Report, MR-8, pp. 1–15.

22. "Apollo 13," NASA Mission Report, MR-7, May 20, 1970. Edgar M. Cortright, editor, *Apollo Expeditions to the Moon,* pp. 247–63.

23. David Baker, *The History of Manned Spaceflight,* pp. 403, 405, 406. Also Max Ary, pers. corr. to L. Purnell, October 8, 1982. And "Apollo 14," NASA Mission Report, n.d. (in NASM Archives). "Apollo 14, Flight To Fra Mauro," NASA Facts,

Johnson Space Center, 1977.

24. "Apollo 17, The Last Apollo," NASA Facts, NASA-JSC, n.d. General Electric Company, news release, April 9, 1971, pp. 1–6. Baker, *History of Manned Spaceflight,* pp. 410–23, 431–35, 437–50. Also Gene Simmons, *On The Moon With Apollo 17,* pp. 1–19.

25. Simmons, *On the Moon with Apollo 17,* pp. 1–19.

26. Baker, *History of Manned Spaceflight,* pp. 437–50.

27. Hamilton Standard, "Mobility Range and Ventilation Study, Apollo Suit A-3H-0234," July 15, 1964. Also see "Apollo Space Suit Development," "Basic SSA Costs," dated 1964, handwritten notes (in NASM Archives). Also Radnofsky/Grimwood interview, March 21, 1966, p. 3.

28. Hamilton Standard, Progress Report for 8/1/64–9/30/64 for Apollo Space Suit System, October 15, 1964. Also R. Jones, "Evaluation and Comparison of Three Space Suit Assemblies," Technical Note NAS TND-3482, July 1966. John Rayfield, ILC Dover, pers. corr., January 2, 1986, and pers. corr. January 2, 1987.

29. Telephone conversation with Joseph Kosmo, Crew Systems, Johnson Space Center, November 19, 1985. Also see William Hines, "Fireproof Fabric Perfected, Use on Apollo Flight Likely," *The Evening Star* (Washington, D.C.), August 18, 1967, p. A-8. Also "NASA To Get Space Suits of Fiberglass," *The Washington Post,* August 18, 1967, p. A-2. Also Don Partner, "Suits for Astronauts," *The Sunday Denver Post,* August 27, 1967. And "New Apollo Space Suit to be Tested in September, *Space Daily,* August 18, 1967, p. 252.

30. "Redesigned Apollo Suits to be Tested," NASA news release, No. 67-222, August 18, 1967. Also see *Hamilton Standard News,* January 23, 1969. And *Familiarization and Operations Manual, Model A7L, Apollo Block II,* April 5, 1968.

Chapter 6 Space Suits for the Manned Orbiting Laboratory, Skylab, and the Apollo-Soyuz Test Project

1. Walter A. McDougall, . . . *the Heavens and the Earth* (New York: Basic Books, Inc., 1985), pp. 419–421. David W. Compton and Charles D. Benson, *Living and Working in Space: A History of Skylab,* pp. 7–15.

2. David Baker, *The History of Manned Space Flight,* pp. 182, 215, 250, 361, 451–64, 507–32.

3. Heather M. David, "New Type of Space Suit May Be Needed for MOL Astronauts," *Missiles and Rockets* (Dec. 7, 1964):22. Also see "NASA, USAF Study Two-Gas Atmospheres," *Aviation Week & Space Technology* (Aug. 15, 1966):79. Also see Aleck C. Bond, memorandum to Director of Engineering and Development, Manned Spacecraft Center, concerning MOL space suit development at Hamilton Standard, August 1, 1969 (in NASM Archives).

4. Compton and Benson, *Living and Working in Space,* p. 353.

5. Bruce Ferguson, Public Relations Director, ILC, "The New Apollo and Skylab Space Suits," n.d.; also see Baker, *The History of Manned Spaceflight,* pp. 490–506.

6. NASA, Johnson Space Center, Crew Systems Div., *Familiarization Manual for Skylab; 28-Day Clothing Module and Contingency Clothing Module,* CSD-S-038 (Houston, Texas: JSC, 1968), pp. 2–6; also see "Space Garments for IVA wear," *Skylab Experience Bulletin #6* (JSC), 1974; and Baker, *The History of Manned Spaceflight,* p. 453.

7. Thomas Y. Canby, "Skylab, Outpost on the Frontier of Space," *National Geographic,* vol. 146, (1974):451–69. Compton and Benson, *Living and Working in Space,* pp. 253–353.

8. Edward Clinton Ezell and Linda Neuman Ezell, *The Partnership: A History of the Apollo-Soyuz Test Project,* pp. 64–65, 80–81, 300, 314, 347, 349, 355–58.

9. NASA Press Release, "Apollo-Soyuz Test Project," 1975, pp. 59–62, 69. Also see S. P. Umanskiy, *Chelovek na Kosmicheskoy Orbite,* pp. 77, 90.

Chapter 7 Space Shuttle Suits

1. "Shuttle Suit to Offer Improved Mobility," *Aviation Week & Space Technology* (April 5, 1976): 59; and "Ham Standard Picked to Build EVA Spacesuits for Shuttle Crews," *Roundup* (Johnson Space Center) 15 (July 30, 1976).

2. Craig Covault, "Shuttle EVA Suits Incorporate Advances," *Aviation Week & Space Technology* (March 16, 1981):69–73. Also Richard C. Wilde, "EMU—A Human Spacecraft," in *Proceedings of the Fourteenth International Symposium on Space Technology and Science,* Tokyo, 1984, pp. 1565–76.

3. Henry S. F. Cooper, Jr., "A Reporter At Large (The Space Shuttle—Part 1)," *The New Yorker* (February 9, 1981):81. Also an interview with John R. Rayfield, ILC Dover, and Richard C. Wilde, Hamilton Standard, December 12, 1986.

4. Craig Covault, "Shuttle Suit Shows Advances on Apollo," *Aviation Week & Space Technology* (August 15, 1977):37–40.

5. Interview with Richard C. Wilde, Hamilton Standard, December 12, 1986. Telephone conversation with David Cadogan, ILC Dover, Inc. September 22, 1992. Trish Donnally, "Aviator Chic," *The Washington Times Magazine,* July 4, 1984, p. 161 (in NASM Archives).

6. Martin Marietta, "Manned Maneuvering Unit," March 1985. Joseph P. Allen, *Entering Space: An Astronaut's Odyssey,* pp. 103–13. Also see Dave Dooling, "Dressing Up for a Walk in Space," *Space World* (October–November 1982):13–15; and Michael A. Sposito, "An Interview with Astronaut Joe Allen," *Astronomy* 14 (November 1986):26–32. In this interview, Allen points out that in order to save time on the prebreathe, the Shuttle pressure had been dropped from 14.7 to 10 psi the day before the EVA, thus cutting prebreathe time from 3 hours to 45 minutes.

7. Allen, *Entering Space,* p. 106. Also pers corr. from Richard C. Wilde, Hamilton Standard, to the author, December 12, 1986.

8. Johnson Space Center, "EVA Systems," *Shuttle Flight Operations Manual* 15 (June 30, 1980):2.2-17 to 2.2-31. Also see "Shuttle Space Suit," NASA Educational Brief, EB-81-2, n.d.; ILC Industries, Inc., "Material Specification for Coated, Nylon, Ripstop, Neoprene Coated (Shuttle Space Suit Fabric)," report 108-1-11F, July 14, 1971; ILC Dover, "Material Specification for Cloth Nylon, Tricot," report 108-1-10C, September 21, 1982; and Midwest Research Institute, "Case Study: Liquid Cooled Garments," August 12, 1974, p. 11.

9. "Gore-Tex Expanded PTFE" [W. L. Gore & Associates Inc.], n.d. In pers. corr. from Laurie Gil to the author.

10. *United Technologies Hamilton Standard News,* "Space Shuttle Space Suit Life Support System," n.d.

11. William R. Pogue, *How Do You Go to the Bathroom in Space?,* pp. 49–50. Also interview with Richard C. Wilde, Hamilton Standard, December 12, 1986. Interview (May 15, 1987) with Duane F. Bueley, a consulting engineer who, together with Dr. Jake Ellis, Don Alford, and James Rivers, devel-

oped the Ellis Fitting into a usable backup system for the DACT.

12. Trish Donnally, "Aviator Chic," *The Washington Times Magazine* (July 24, 1984) (in NASM Archives). Also see Allen, *Entering Space,* p. 191; and Eloise Engle and Arnold S. Lott, *Man in Flight,* pp. 206–07. Nigel Macknight, *Shuttle,* pp. 50–51. Also see Howard Benedict, *NASA: A Quarter Century of Space Achievement,* pp. 250–51.

13. Allen, *Entering Space,* pp. 119, 203–36. See George Torres, *Space Shuttle A Quantum Leap,* pp. 48–69. See Andrew Wilson, *Space Shuttle Story,* pp. 47–115. Also NASA/Lewis, *Space Shuttle Activities for Primary and Intermediate Students* (Cleveland, Ohio: NASA Lewis, 1987), pp. 1.23–1.26. Also Ann Davis Nunemacher, "Seamstresses Practice Space Age Sewing for NASA," *Sew News,* December 1986, pp. 30, 31.

14. John Grossmann, "The Blue Collar Spacesuit," *Air & Space,* October/November 1989, vol. 4, no. 4, p. 62. Kathryn Sullivan, "Walk-in-Space," Smithsonian National Associates Program lecture, March 2, 1989.

15. Craig Covault, "Atlantis' Radar Satellite Payload Opens New Reconnaissance Era," *Aviation Week & Space Technology,* vol. 129, no. 24, December 12, 1988, pp. 26–28. See also, "The Magic is Back!", *Time,* vol. 132, no. 15, pp. 20–25. Tim Furniss, *Space Shuttle Log* (New York: Jane's Publishing Inc., 1986), pp. 48–64. James R. Asker/Houston, "U.S. EVAs Resume As Astronauts Help Deploy Satellite," *Aviation Week & Space Technology,* vol. 134, no. 18, May 6, 1991, pp. 18–20, 21, 22, 42–45.

16. Bruce W. Sauser, Project Engineer, "Space Shuttle Phase I Crew Escape Equipment," NASA Johnson Space Center, Houston, Texas, n.d. Walter A. McDougall, . . . *the Heavens and the Earth* (New York: Basic Books, Inc., 1985), p. 434. Craig Covault, "Japan Forging Aggressive Space Development Pace," *Aviation Week & Space Technology,* vol. 133, no. 7, pp. 36–39.

Chapter 8   Advanced Development Space Suits

1. Matthew I. Radnofsky, interview by James M. Grimwood, March 21, 1966. NASA Headquarters, "Space Station Appointments Announced at Johnson Space Center," release 84-50, April 9, 1984.

2. Litton Systems, Inc., "The Litton Extra-

vehicular and Lunar Surface Suit," publication 6826, ATD 11-67-004, Dec. 1967, p. 1. Radnofsky interview.

3. "Space Flight—Right in the Lab," reprint of article from *Business Week,* October 12, 1957. "Laboratory Simulates 95-Mi. Altitude," reprint from *Aviation Week,* October 14, 1957. "Vacuum Chamber Helping Man Solve Problems of Outer Space," Associated Press article in Siegfried Hansen's personal scrapbook, no newspaper name, January 25, 1958. William B. Harris, "Litton Shoots for the Moon," *Fortune,* April 1958, pp. 114–16. Siegfried Hansen, pers. comm. to author, October 19, 1990. I am grateful to Dennis Gilliam for bringing the article "Fit, Flexibility, Heat Control are Vital in Space Suit," by Russell Hawkes, *Aviation Week,* June 22, 1959, to my attention.

4. Heather M. David, "NASA Considering Litton-Developed 'Hard Shell' Suit for Explorations," *Missiles and Rockets* (Aug. 24, 1964):22.

5. Litton Systems, "The Litton Extravehicular and Lunar Surface Suit," pp. 2–6. David, "NASA Considering," p. 23.

6. C. M. Plattner, "Advanced Space Suits Developed by Litton," *Aviation Week & Space Technology* (April 14, 1969):71–76. Johnson Space Center, "Advanced Development Suit Programs," n.d. (synopsis of programs from 1962 through 1974) (in NASM Archives).

7. M. C. Vykukal and B. W. Webbon, "High-Pressure Protective Systems Technology," paper, 9th Intersociety Conference on Environmental Systems, San Francisco, Calif., July 16–19, 1979, p. 2, sponsored by the Aerospace Division of the American Society of Mechanical Engineers (in NASM Archives).

8. John Thiel, unpublished notes, n.d. (in NASM archives).

9. Tom Alexander, "The Man in the Moon Suit," reprinted by permission of *Esquire,* 1963, for Hamilton Standard. Vykukal and Webbon, "High-Pressure Protective Systems," p. 3.

10. Stacy V. Jones, "Patents Protective Joints for Space Suit," *The New York Times,* July 12, 1986 (in NASM Archives). Michael Mecham, "Space Suit Design Gets Hard Look," *USA Today,* July 22, 1986.

11. C. W. Flugel, J. J. Kosmo, and J. Rayfield, "Development of a Zero-PreBreathe Spacesuit," paper, 14th Intersociety Conference on Environmental Systems, San Diego, Calif., July, 16–19, 1984. And ILC presentation handout, "Advanced Spacesuit Technology Programs at ILC Dover, Inc., December,

1986. John Grossmann, "The Blue Collar Spacesuit," *Air & Space,* October/November 1989, vol. 4, no. 4, pp. 58–67. Carol A. Shifrin, "NASA to Evaluate Two Suit Designs for Space Station," *Aviation Week & Space Technology,* January 11, 1988, pp. 36–39. Joseph J. Kosmo and William E. Spenny, NASA, Johnson Space Center; Rob Gray and Phil Spampinato, ILC/Dover, Inc., "Development of the NASA ZPS Mark III 57.2-kN/m² (8.3 psi) Space Suit," paper, 18th Intersociety Conference on Environmental Systems, San Francisco, Calif., July 11–13, 1988. J. L. Zelon, ILC Space Systems, Houston, Texas, "Advanced Space Suit Development for Future On-Orbit Operations, paper, AIAA Space Station in the Twenty-First Century, September 3–5, 1986, Reno, Nevada. Phil Spampinato, David Cadogan, Tony McKee, ILC Dover, Inc.; Joseph J. Kosmo, NASA, Johnson Space Center, "Advanced Technology Application in the Production of Spacesuit Gloves," # 901332.

Chapter 9 Space Suits in the National Collection

1. Joseph D. Atkinson, Jr., and Jay M. Shafritz, *The Real Stuff,* pp. 97, 98. Interview with Louis R. Purnell, first curator of the space suit collection, and other manned spaceflight artifacts, October 17, 1984.

2. Interview with Gregory P. Kennedy, former curator of the manned spaceflight collection, October 18, 1984. "Space Suits for Project Apollo," ILC Industries, Inc., 1969.

3. Brooke Hindle, "How Much is a Piece of the True Cross Worth?" in *Material Culture and the Study of American Life,* Ian M. G. Quimby, ed. (New York: W. W. Norton Company, 1978).

4. R. M. Organ, "Interim Report of Space Suits," Conservation Analytical Laboratory, Smithsonian Institution, 1978.

5. Sharon Blank, "Basic Care and Conservation of Synthetic Materials," lecture given for National Air and Space Museum's Conservation Seminar, 1987. Also Mary T. Baker, "Spacesuit Samples," CAL report, June 15, 1990.

6. Mary Ballard and Tim Padfield, "Amended Report on Storage and Display of Space Suits," Conservation Analytical Laboratory, Smithsonian Institution, 1987.

7. John S. Mills and Raymond White, *The Organic Chemistry of Museum Objects,* pp. 64, 113, 114.

8. Virginia Lee Pledger, presentation for NASM Exhibits Department, April 1989. (Textile conservator in private practice, 621 East Capitol Street, S. E., Washington, D.C. 20003, telephone 202-543-2510.)

9. Mary T. Baker, "Spacesuit Samples." Also Ed McManus, "Summary of 15 June Meeting Concerning the Status of the NASM Space Suit Collection," memo, June 19, 1990. Garry Thomson, *The Museum Environment,* pp. 20, 21. Ed McManus, memo to Ken Isbell, of the Space History Department, February 6, 1990. Sylvia C. Fishman, *Stumpwork Society Chronicle,* February 1990, vol. 12, no. 1, p. 2.

10. Mary T. Baker and Ed McManus, "Space Suits: NASA's Dream—Conservator's Nightmare," report, November 1991.

11. Parts of this chapter were originally presented as a paper at the Eighth Annual Conference on Textiles, August 1989, at the University of Maryland, and published in the *Ars Textrina Journal,* vol. 11 (1989):77–106.

Select Bibliography

Books and Book-length Publications

Allen, Joseph P., with Martin, Russell. *Entering Space: An Astronaut's Odyssey.* New York: Stewart, Tabori & Chang Pub., 1984.

Atkinson, Joseph D., Jr., and Shafritz, Jay M. *The Real Stuff.* New York: Praeger Publishers, 1985.

Baker, David. *The History of Manned Spaceflight.* New York: Crown Publishers, 1981.

Benedict, Howard. *NASA: A Quarter Century of Space Achievement.* The Woodlands, Texas: Pioneer Publications, 1984.

Brooks, Courtney G.; Grimwood, James M.; and Swenson, Loyd S., Jr. *Chariots for Apollo.* Washington, D.C.: NASA, 1979.

Caidin, Martin, and Caidin, Grace. *Aviation and Space Medicine.* New York: E. P. Dutton, 1962.

Collins, Michael. *Carrying the Fire.* New York: Bantam Books, 1974.

Compton, W. David, and Benson, Charles D. *Living and Working in Space: A History of Skylab,* NASA SP-408. Washington, D.C.: NASA, 1983.

Cortright, Edgar M., ed. *Apollo Expeditions to the Moon,* NASA SP-350. Washington, D.C.: NASA, 1975.

Crouch, Tom. *The Bishop's Boys.* New York: W. W. Norton Company, 1989.

———. *The Eagle Aloft.* Washington, D.C.: Smithsonian Institution Press, 1983.

Dille, John, ed. *We Seven.* New York: Simon & Schuster, 1962.

Engle, Eloise K., and Lott, Arnold S. *Man in Flight.* Annapolis, Md.: Leeward Publications, 1979.

Ezell, Edward Clinton, and Ezell, Linda Neuman. *The Partnership: A History of the Apollo-Soyuz Test Project,* NASA SP-409. Washington, D.C.: NASA, 1978.

Grimwood, James M.; Hacker, Barton C.; with Vorzimmer, Peter J. *Project Gemini, A Chronology.* Washington, D.C.: NASA, 1969.

Haldane, John Scott, and Priestly, J. G. *Respiration.* New Haven, Conn. and London: Yale University Press, 1935.

Hartfield, John W. *Space Shuttle Activities for Primary and Intermediate Students.* Cleveland, Ohio: NASA, 1987.

Hindle, Brooke. "How Much is a Piece of the True Cross Worth?" In *Material Culture and the Study of American Life,* edited by Ian M. G. Quimby. New York: W. W. Norton Company, 1978.

Johnson, Richard F.; Correale, Richard S.; and Radnofsky, Matthew I. *Space Suit Development, NASA TN D-3291.* Washington, D.C.: NASA, 1966.

Johnson, Richard F.; Dietlein, Lawrence F., M.D.; and Berry, Charles A., M.D., eds. *Biomedical Results of Apollo, NASA SP-368.* Washington, D.C.: NASA, 1975.

Machell, R. M. et al. "Crew Station and Extravehicular Equipment." In *Gemini Midprogram Conference,* February 23–25, 1966, NASA-SP-121. Washington, D.C.: NASA, 1966.

Macknight, Nigel. *Shuttle.* Osceola, Wisc. Motorbooks International, 1985.

Macready, John A. "Airplane Reaches Altitude of 7.75 Miles." In *Test Pilots,* edited by Gene Gurney. New York: Arno Press, 1980.

Mallan, Lloyd. *Suiting Up for Space.* New York: The John Day Co., 1971.

Mills, John S., and White, Raymond. *The Organic Chemistry of Museum Objects.* London: Butterworth & Co., Ltd., 1987.

Mohler, Stanley R., and Johnson, Bobby H. *Wiley Post, His Winnie Mae, and the World's First Pressure Suit.* Washington, D.C.: Smithsonian Institution Press, 1971.

Newell, Homer E. *Beyond the Atmosphere, NASA SP-4211.* Washington, D.C.: U.S. Government Printing Office, 1980.

Pogue, William R. *How Do You Go to the Bathroom in Space?* New York: Tom Doherty Assoc., 1985.

Simmons, Gene. *On The Moon With Apollo 17, NASA EP-101.* Washington, D.C.: NASA, 1972.

Still, E. W. *Into Thin Air.* London: Sydenham and Co., 1958.

Stine, G. Harry. *The Handbook for Space Colonists.* New York: Holt, Rinehart and Winston, 1985.

Sweeting, C. G. *Combat Flying Clothing.* Washington, D.C.: Smithsonian Institution Press, 1984.

———. *Combat Flying Equipment.* Washington, D.C.: Smithsonian Institution Press, 1989.

Swenson, Loyd S., Jr.; Grimwood, James M.; and Alexander, Charles C. *This New Ocean: A History of Project Mercury.* Washington, D.C.: NASA, 1966.

Taylor, John W. R. *Air Facts and Feats.* London: Guiness, 1973.

Thomson, Garry. *The Museum Environment.* London: Butterworth & Co., Ltd., 1986.

Torres, George. *Space Shuttle A Quantum Leap.* Novato, Calif.: Presidio Press, 1985.

Umanskiy, S. P. *Chelovek na Kosmicheskoy Orbite* (Man in Space Orbit), NASA TT F-15973, translation. Moscow: Mashinostroyeniye Press, 1974.

United Press International. *Gemini.* New Jersey: Prentice-Hall, 1965.

Wilde, Richard C. "EMU—A Human Spacecraft." In *Proceedings of the Fourteenth International Symposium on Space Technology and Science,* Tokyo, 1984, pp. 1565–76.

Wilson, Andrew. *Space Shuttle Story.* London: Hamlyn Publishing, 1986.

Periodicals

"Apollo 9 Verifies Lunar Suit, Backpack." *Aviation Week & Space Technology,* March 17, 1969, p. 67.

"Apollo Space Suit Contract Issued." *Aviation Week & Space Technology,* October 22, 1962, p. 33.

Canby, Thomas Y. "Skylab, Outpost on the

Frontier of Space." *National Geographic* 146 (1974): 451–503.

Cooper, Henry S. F., Jr. "A Reporter At Large (The Space Shuttle-Part 1)." *The New Yorker,* February 9, 1981, p. 81.

Correale, James V., and Guy, Walter W. "Space Suits," *Space World,* September-October 1963, p. 16.

Covault, Craig. "Discovery Crew Deploys TDRS, Test Space Station System." *Aviation Week & Space Technology,* March 20, 1989.

———. "Galileo Launch to Jupiter by Atlantis Culminates Difficult Effort With Shuttle." *Aviation Week & Space Technology,* October 2, 1989, pp. 54–67.

———. "Shuttle EVA Suits Incorporate Advances." *Aviation Week & Space Technology,* March 16, 1981, pp. 69–73.

———. "Shuttle Launch of Galileo Jupiter Mission Highlights U.S. Space Science Renaissance." *Aviation Week & Space Technology,* October 23, 1989, pp. 35, 36.

———. "Shuttle Launch Schedule Accelerates After Galileo Deployment, Atlantis Recovery." *Aviation Week & Space Technology,* October 30, 1989, pp. 22, 23.

———. "Shuttle Suit Shows Advances on Apollo." *Aviation Week & Space Technology,* August 15, 1977, pp. 37–40.

David, Heather M. "Apollo Suit Substantially Redesigned." *Missiles and Rockets,* April 26, 1965, pp. 26–27.

———. "NASA Considering Litton-Developed 'Hard Shell' Suit for Explorations." *Missiles and Rockets,* August 25, 1964, p. 22.

———. "New Type of Space Suit May Be Needed for MOL Astronauts." *Missiles and Rockets,* December 7, 1964, p.22.

Dooling, Dave. "Dressing Up for a Walk in Space." *Space World,* October-November 1982, pp. 13–15.

Flinn, John. "Manchester's Velcro Celebrates 40 Years of Stick-To-Itiveness, Fastener's Rrrrrrrip Can be Heard Everywhere." *The Union Leader,* Manchester, N.H., December 9, 1988.

Friedel, Robert. "The First Plastic, Invented in 1869, It Was an Elegant, Pioneering Cheap Substitute for Finer Things." *Invention and Technology,* Summer, 1987, vol. 3, no. 1, pp. 18–23.

Gell, Charles F.; Hays, Edward L.; and Correale, James V. "Development History of the Aviator's Full Pressure Suit in the U.S. Navy." *The Journal of Aviation Medicine* 30 (1959): pp. 242–48.

Harris, William B. "Litton Shoots for the Moon." *Fortune,* April 1958, pp. 114–16.

Hines, William. "Fireproof Fabric Perfected, Use on Apollo Flight Likely." *The Evening Star* (Washington, D.C.), August 18, 1967, p. A-8.

Holder, Dennis. "Sticky Business." *Philip Morris Magazine,* Fall, 1988, vol. 3, no. 4, pp. 27–29.

Hounshell, David A., and Smith, John Kenly, Jr. "The Nylon Drama." *Invention and Technology,* Fall 1988, vol. 4, no. 2, pp. 40–48.

Keovsk, William T. "ILC's customers include the men on the moon." *The Sunday Bulletin* (Philadelphia), October 18, 1970, *Prospect '71,* p. 43.

Lasagna, Louis. "With a Wing and a Prayer." *The Sciences,* March-April 1985, p. 8.

Miles, Marvin. "Men 'Soar' 89 Miles—on Ground." *Los Angeles Times,* October 7, 1957, p. 14.

"NASA, USAF Study Two-Gas Atmospheres." *Aviation Week & Space Technology,* August 15, 1966, p. 79.

"NASA To Get Space Suits of Fiberglass." *The Washington Post,* August 18, 1967, p. A-2.

"New Apollo Space Suit to be Tested in September." *Space Daily,* August 18, 1967, p. 252.

Pavlik, Rick, "Next Space Launch Scheduled November 17." *Spacewatch,* November 1988, vol. 5, no. 11.

Plattner, C. M. "Advanced Space Suits Developed by Litton." *Aviation Week & Space Technology,* April 14, 1969, pp. 71–76.

Post, Wiley. "Wiley Post." *Popular Mechanics,* October 1934, pp. 492–95.

Sawyer, Kathy, and Peterson, Cass. "Americans Soar Back to Space Aboard Discovery, Program's Long Grounding Ends; Shuttle's Crew Deploys Satellite." *The Washington Post,* September 30, 1988, pp. 1, 12.

Schirra, Walter. "A Suit Tailor-Made for Space." *Life,* August 1, 1960, p. 36.

"Shuttle Suit to Offer Improved Mobility." *Aviation Week & Space Technology,* April 5, 1976, p. 59.

Smith, Randy. "Post and Rogers—Se-Tenants in Life." *Stamp World,* August 1983, p. 62.

"Space Flight—Right in the Lab." *Business Week,* October 12, 1957, pp. 78–81.

Sposito, Michael A. "An Interview with Astronaut Joe Allen." *Astronomy,* November 1986, vol. 14, no. 6, p. 3.

Stone, Irving. "Laboratory Simulates 95-Mi. Altitude." Reprint from *Aviation Week,* October 14, 1957.

"USAF Full Pressure Space Suit is Light, Permits Free Movement." *Aviation Week,* December 9, 1957, p. 29.

von Braun, Wernher. "What an Astronaut Will Wear on the Moon." *Popular Science,* April, 1965, pp. 87–88.

Walker, T. W. "The Development of the Pressure Suit for High Altitude Flying." *The Project Engineer,* May 1956, vol. 15, p. 4.

"Water-Cooled Apollo Suit Prototype Shipped to NASA." *Aviation Week,* March 23, 1964.

"Water-Cooled Undies." *Mechanix Illustrated,* December 1964, p. 85.

Wetmore, Warren G. "Improved Suit Proposed for Lunar Wear." *Aviation Week & Space Technology,* March 3, 1969, p. 58.

Technical Documents

David Clark Company. "Field Maintenance Instructions for Gemini Space Suit Type G-3C." October 1964.

———. "Gemini EVA Space Suit Assembly Familiarization Handbook." n.d. pp. 4–26.

———. "Report of New Technology, Project Gemini Space Suit Program." May 11, 1964.

———. "Statement of Work for Hazardous Environment Protective Assembly (S-1C)."

———. "Technical Manual Maintenance, Repair and Testing of Space Suit Types G-2C." n.d.

E. I. du Pont de Nemours, Inc. "Dupont Kapton Polyimide Film." May 1, 1966.

———. "Nylon's Golden Anniversary," Background News, External Affairs Department, Wilmington, Delaware, January 1988. Material prepared especially for the golden anniversary celebration of nylon (i.e., "Nylon: The Beginning of The Materials Revolution").

———. "Shaping the Future: 50th Anniversary of Nylon," Handouts from A Lifestyle Show presented by the E. I. du Pont Company and Mazza Gallerie, Washington, D.C., January 1988.

———. "Summary of Properties, Dupont H Film polyimide film," Technical Information Bulletin H-1.

———. "Summary of Properties, Dupont Mylar Polyester," Technical Information Bulletin M-1C.

Flugel, C. W.; Kosmo, J. J.; and Rayfield, J. "Development of a Zero-PreBreathe Spacesuit." Paper, 14th Intersociety Conference on Environmental Systems, San Diego, Calif., July 16–19, 1984.

General Electric Company. News Release, April 9, 1971, pp. 1–6.

Hamilton Standard. "Mobility Range and Ventilation Study, Apollo Suit A-3H-0234." July 15, 1964.

———. "Performance and Design Requirements for the Extra-Vehicular Mobility Unit." Apollo Program General Specifications. Windsor Locks, Conn.: Hamilton Standard, 1965. pp. 3.0–3.1.2.

———. Progress Report for 8/1/64–9/30/64 for Apollo Space Suit System. October 15, 1964.

———. "Space Shuttle Space Suit Life Support System." *United Technologies Hamilton Standard News.* n.d.

ILC Industries. *Familiarization and Operations Manual, Model A7L, Apollo Block II.* April 5, 1968.

———. "Material Specifications for Cloth, Nomex, 6 oz." International Latex Corporation, August 25, 1966.

———. "Material Specification for Cloth Nylon, Tricot." ILC Dover, report 108-1-10C, September 21, 1982.

———. "Material Specification for Cloth, Polyester, Non-woven." ILC Dover, July 13, 1971.

———. "Material Specification for Cloth, Ripstop Nylon, Neoprene Coated, 3.5 oz." ILC Dover, March 29, 1968.

———. "Material Specification for Coated, Nylon, Ripstop, Neoprene Coated (Shuttle Space Suit Fabric)." ILC Industries, report 108-1-11F, July 14, 1971.

———. "Material Specifications for Nylon Spandex, 5 and 6 ounces." ILC Dover, December 9, 1966.

———. "Material Specification for Tubing, Plastic, Flexible, Modified PVC." ILC, Dover, November 2, 1966.

———. "Space Suits for Project Apollo." ILC Industries, 1969.

———. "Spacesuit Technology Programs at ILC Dover, Inc." ILC Dover, December 1986.

Litton Systems. "The Litton Extravehicular and Lunar Surface Suit." Publication 6826, ATD 11-67-004, December 1967, p. 1.

Lompart, H.; Herrington, L. P.; and Hoff, E. C. "Review of Methods of Applying Air Pressure to the Extremities for Protection Against Acceleration with Measurement of the Effective Pressure on the Skin." CAM Report #228, OSRD, November 14, 1943.

Marcum, A. L., and Mauch, H. A. "Experimental Space Worker's Garment and Helmet Assembly." Technical Documentary Report No. AMRL-TDR-64-37, May 1964, pp. 4–13.

Martin Marietta. "Manned Maneuvering Unit." March 1985.

Midwest Research Institute, "Case Study: Liquid Cooled Garments," August 1, 1974, p. 11.

Roth, J. A. "Report on Pressure Altitude Suit Equipment," *AAF Proving Ground Command Report,* July 13, 1943, AAF Board Project No. (M-4) 314.

Schulz, Helen W. "Case History of Pressure Suits." Report prepared by the Historical Office Executive Secretariat, Air Materiel Command, Wright-Patterson Air Force Base, May 1951, p. 19 (Digest).

"Space Shuttle Return to Flight," pamphlet prepared by Honeywell, Martin Marietta, Morton Thiokol, Inc., Rockwell International (Rocketdyne and Space Transportation Systems Division, United Technologies USBI) n.d.

U.S. Department of the Navy, Bureau of Aeronautics. "Buaer Aviation Clothing and Survival Equipment." Bulletin No. 1-59. AER-AE-5111/25, June 18, 1959. p. 3.

U.S. Department of the Navy, Bureau of Naval Personnel. "Parachute Rigger 1 & C." Navy Training Courses, NAVPERS 10360. Washington, D.C.: GPO, 1959.

Vykukal, M. C., and Webbon, B. W. "High-Pressure Protective Systems Technology." Paper, 9th Intersociety Conference on Environmental Systems, San Francisco, Calif., July 16–19, 1979, p. 2, sponsored by the Aerospace Div. of the American Society of Mechanical Engineers.

National Aeronautics and Space Administration Technical Documents

NASA Headquarters. "Space Station Appointments Announced at Johnson Center." Release 84-50, April 9, 1984.

Johnson Space Center. "Advanced Development Suit Programs." n.d. Synopsis of programs from 1962 through 1974.

———. "Apollo 14, Flight To Fra Mauro." NASA Facts, 1977.

———. "Apollo-Soyuz Test Project." Press Release, 1975.

———. "EVA Systems." *Shuttle Flight* Operations Manual 15 (June 30, 1980): .−17 to .−31.

———. "Shuttle Space Suit." NASA Educational Brief, EB-81, n.d.

———. "Space Garments for IVA wear." Skylab Experience Bulletin 6, 1974.

———. "STS-29: Completion of the TDRSS Constellation." NASA Activities, vol. 20, no. 3, March 1989, p. 8.

———. "STS-29." NASA Report to Educators, vol. 17, no. 1, Spring 1989.

———. "U.S. Manned Space-Flight Log," Astronaut Fact Book, NASA Information Summaries, PMS-011A(JSC), March 1988, p. 39.

Manned Spacecraft Center. "Apollo 9." NASA Mission Report, MR-3. March 26, 1969.

———. "Apollo 12." NASA Mission Report, MR-8, pp. 1–15.

———. "Apollo 13." NASA Mission Report, MR-7, May 20, 1970.

———. "Apollo 14." NASA Mission Report, n.d.

———. "Apollo 17, The Last Apollo." NASA Facts, n.d.

———. "Apollo Extravehicular Mobility Unit Design and Performance Specification." MSC-CSD-A-017, December 28, 1964. p. 3–102.

———, Barker, J. T. "Project Gemini Extra-Vehicular Pressurization Ventilation System Statement of Work." Crew Equipment Branch, June 19, 1963, pp. 1–2.

———, Clark, David M. "On Partial Pressure Suits." In a space suit history compiled by C. C. Lutz, Crew Systems Div., pp. 3–12.

———, Crew Systems Division. *Familiarization Manual for Skylab: 8-Day Clothing Module and Contingency Clothing Module.* CSD-S-038. Houston, Texas: Johnson Space Center, 1968.

———, Crew Systems Division. "Design Certification Report on G-5C Space Suit for GT-7 Mission." n.d.

———, Ertel, Ivan D. "Gemini Program." Fact Sheet 291-F, August 1966.

———. "Field Maintenance Instructions for Gemini Space Suit Type G-3C." April/October 1964.

———. "Gemini 4 Flight." Fact Sheet 291-B, 1965.

———. "Gemini VII/Gemini VI." Fact Sheet 291-D, January 1966.

———. "Gemini VIII Rendezvous and Docking Mission." Fact Sheet 291-E, April 1966.

———. "Gemini IX Flight Suit History and Configuration." Flight Readiness Review, May 2, 1966.

———. "Gemini IX-A Rendezvous Mission." Fact Sheet 291-F, August 1966.

———. "Gemini X Flight Suit History and Configuration Flight Readiness Review," July 5, 1966.

———. "Gemini X, Multiple Rendezvous, EVA Mission." Fact Sheet 291-G, September 1966.

———. "Gemini XI Mission, High Altitude, Tethered Flight." Fact Sheet 291-H, October, 1966.

———, Johnston, Richard S., and Hays, Edward L. "The Development and Operation of Extravehicular Equipment." 1966.

———, Jones, R. "Evaluation and Comparison of Three Space Suit Assemblies." Technical Note NASA TND-3482, July, 1966.

———. "NASA Certification Review, GT-3, Gemini Intravehicular Space Suit Assembly." February 1965.

———. "NASA to Negotiate for Apollo Suit, Support System." News Release 65-346, November 5, 1965.

———. News Release 63-110, July 5, 1963.

———. News Release. "NASA Contract to FRL." Fabric Research Laboratories, May 3, 1965.

———. "Pressure Garment Assemblies required for NAA," S&ID Apollo Test Program, Enclosure I, n.d.

———. "Project Gemini, Extravehicular Pressurization Ventilation System, Statement of Work." June 19, 1963.

———. "Redesigned Apollo Suits to be Tested." News Release 67-222, August 18, 1967.

———, Spross, F. R. "Gemini XII Postflight

Space Suit Evaluation." November 23, 1966.

———. "Statement of Work, Gemini Space Suit Reliability Testing Program." December 2, 1963.

———, Tucker, E. M. "Gemini X Postflight Space Suit Evaluation." Gemini Space Suit Office, August 2, 1966.

———, Tucker, E. M. "GT-6 Post-Flight Report, Space Suits and Accessories." January 4, 1966.

———, Tucker, E. M. "GT-7 Space Suit Post Flight Report." January 4, 1966.

References Available in the Archives of the National Air and Space Museum

Air Force Physiological Training Program News Letter, Number 44, June 1961, p. 5.

Alexander, Tom. "The Man in the Moon Suit." Reprinted by permission of *Esquire,* 1963, for Hamilton Standard, Division of United Aircraft Corporation.

"Astronauts Show Off New Fireproof Space Suit." *The New York Times,* August 3, 1978.

Burkett, Warren. "Eight Gemini Astronaut Suits Begun." *Missiles and Rockets,* May 25, 1964.

Colley, Russell. Oral history interviews, May 9–11, 1989.

Donnally, Trish. "Aviator Chic." *The Washington Times Magazine,* July 4, 1984.

Eberhart, Jonathan. "U.S. Astronauts Could Fly without Space Suits." Science Service News Release, October 13, 1964.

Ferguson, Bruce, Public Relations Director, ILC Industries. "The New Apollo and Skylab Space Suits." n.d.

"Ham-Standard Chosen to Build Shuttle Suits." *Roundup 15,* Johnson Space Center, July 30, 1976.

Jones, Stacy V. "Patents Protective Joints for Space Suit." *The New York Times,* July 12, 1986.

Mecham, Michael. "Space Suit Design Gets Hard Look." *USA Today,* July 22, 1986.

Neal, J. Allen. "High Altitude Pressure Suits." In *History of Wright Air Development Center, July 1–December 31, 1955.* Vol. IV. unpublished, pp. 27–31.

Partner, Don. "Suits for Astronauts." *The Sunday Denver Post,* August 27, 1967.

Index

Entries suffixed by an f denote citations within figure captions; those suffixed by an n, within endnotes; those suffixed by an s, within sidebars; those suffixed by a t, within tables.

2–A full pressure suit, 31

ABMA. *See* Army Ballistic Missile Agency

A1C Apollo space suit, 75–79, 81s, 98s

Advanced Apollo Program. *See* Skylab

Advanced development space suits, 145–68; early history of, 166–67t; zero-prebreathe, 158–64. *See also* Hard suits

Aerobee Rocket, 24s

Aero Medical Laboratory, 16, 22, 26, 28; Apollo space suits and, 85; full pressure suits and, 31–33; partial pressure suits and, 29–30

Aerospace Medicine (Armstrong), 8n

Agena target vehicle, 67, 69

A1H Apollo space suit, 98s, 100f

A2H Apollo space suit, 98s, 101f, 102f

A3H Apollo space suit, 98s, 102f

A4H Apollo space suit, 98s, 103f

A5H Apollo space suit, 81s, 91n, 98–99s

Air emboli, 7

AiResearch Manufacturing, 57, 124

Air Force, U.S., 42; flying suits and, 19; full pressure suits and, 33, 34, 39; Gemini space suits and, 52; hard suits and, 147; MOL space suits and, 110–11; partial pressure suits and, 26–27, 29, 30–31

Air-Lock, Inc., 158

Air pressure, 4, 7, 8, 9

Air Research and Development Command (ARDC), 28, 31, 33, 42

A2L Apollo space suit, 99s, 103f

A3L Apollo space suit, 99s, 104f

A4L Apollo space suit, 99s, 104f

A5L Apollo space suit, 99s

A6L Apollo space suit, 81s, 99s, 106f

A7L Apollo space suit, 81s, 93f, 94t, 96, 99s, 106f

A7LB Apollo space suit, 85s, 99s, 107f, 108t; for Skylab, 112, 112f, 119, 120f

Aldrin, Buzz, 68, 69, 92–93

Allen, Iona, 135f

Allen, Joe, 142

ALSA. *See* Astronaut life support assembly

ALSEP. *See* Apollo Lunar Surface Experiments Package

Altitude, 2, 8, 9, 15, 31

Altitude chambers, 16, 26

Altitude sickness, 8

Ames Research Center (NASA), 155, 157

AMUs. *See* Astronaut maneuvering units

Anders, William, 93f